Carpentry

Volume 3

Learning Resource Centre
Hadlow College
Hadlow
Tonbridge
Kent TN11 0AL
Tel: **01732 853245**

Date of Return	Date of Return	Date of Return

Please note that fines will be charged if this book is returned late

D0303287

Carpentry and Joinery

Volume 3

Edition

Brian Porter

LCG, FIOC, Cert Ed
Former Lecturer at Leeds College of Building

and

Christopher Tooke

LCG, FIOC, Cert Ed
Former Lecturer and Programme Manager at
Ealing, Hammersmith & West London College,
Chief Examiner for the Institute of Carpenters

AMSTERDAM BOSTON HEIDELBERG LONDON NE
PARIS SAN DIEGO SAN FRANCISCO SINGAPORE S

Butterworth-Heinemann is an imprint of Elsevier
Linacre House, Jordan Hill, Oxford, OX2 8DP
30 Corporate Drive, Burlington, MA 01803

First published by Arnold 1986
Second edition, 1991
Reprinted 1999, 2001, 2003, 2004 (twice), 2005
Third edition, 2007

Notice
No responsibility is assumed by the publisher for any injury and/or damage to persons or property as a matter of products liability, negligence or otherwise, or from any use or operation of any methods, products, instructions or ideas contained in the material herein. Because of rapid advances in the medical sciences, in particular, independent verification of diagnoses and drug dosages should be made

British Library Cataloguing in Publication Data
A catalogue record for this book is available from the British Library

Library of Congress Cataloging-in-Publication Data
A catalog record for this book is available from the Library of Congress

ISBN 13: 978-0-7506-6505-6
ISBN 10: 0-7506-6505-X

For information on all Butterworth-Heinemann publications visit our web site at http://books.elsevier.com

Typeset by CharonTec Ltd (A Macmillan Company), Chennai, India
www.charontec.com
Printed and bound in Great Britain

07 08 09 10 11   10 9 8 7 6 5 4 3 2 1

Contents

Foreword

by David R Winson, Registrar of the Institute of Carpenters

As with previous books in this series, anybody having an interest in carpentry and joinery, whether they are undertaking basic study, additional study to prepare for an Institute of Carpenters advanced qualification, or even purely as a reference work, will find this book an essential and invaluable addition to their expanding toolkit.

This book, the last in a series of three, completes a superb set of reference works for carpenters and joiners. It provides the reader with step-by-step practical advice and sufficient knowledge to reach competency in a number of areas from prefabricated buildings, through stair construction and on to repairs and maintenance of existing buildings. Its text is concisely and factually written, the accompanying illustrations are clearly defined, and where calculations are necessary they are explained in an easy to understand manner, helping the reader to quickly come to terms with the subject matter.

Throughout its pages the book gives unequivocal guidance, particularly in regard to the Building Regulations, Health and Safety and other equally important regulations, enabling the reader to carry out a wide range of different work competently but above all else, safely.

It has again been a privilege to be able to review this revised and updated edition of a much renowned textbook used extensively by those engaged in teaching the craft of carpentry and joinery. This book, along with the other two books of the series, has been included in the Institute of Carpenters' recommended reading list for nearly two decades and will continue to be so.

It has been particularly so in the knowledge that the authors, both of whom I have the privilege of working with, are members of the Institute with a great many years' experience, and membership, to call upon.

I wish all readers every success in their chosen careers and I am confident that this book will, like the others in the series, assist greatly in that success.

Foreword

by Chris Addison, Technical Officer of the British Woodworking Federation

It is a pleasure, on behalf of the British Woodworking Federation, to welcome a new edition of an established textbook on Carpentry and Joinery. Since the original publication twenty years ago, this series has become a classic companion for both the student and the working carpenter and joiner.

Acknowledged experts Brian Porter and Christopher Tooke have completely revised this volume, adding new material and bringing it fully up to date for the twenty-first century. We particularly welcome the extended coverage on the use of fire doors and frames, as part of passive fire protection.

It is always a challenge to keep up to date with Building Regulations and the industry standards whether British, European or international, and the useful end-of-chapter references direct the reader to the relevant information.

From the standpoint of the timber and woodworking industry, the BWF wishes this valuable publication every success.

Preface

The Building Industry is constantly changing with the introduction of new ideas and innovations. Since the last edition of Book 3 we have witnessed numerous changes to the Building Regulations and a movement to bring British Standards in line with Europe. Further pieces of legislation have also had a marked effect on our trade and how we operate – particularly in regard to Health and Safety.

Changes in the type, and sourcing of our main raw material 'wood' in whatever form it takes – be it timber or manufactured boards, is an issue we cannot ignore. Many tropical hardwoods that were once freely available are now more difficult to obtain.

Hardwood and Softwood now come from sustainable sources that use proper forest management. Before wood in its many forms is exported to the UK, it is certificated by one of the following organizations and marked accordingly. This mark identifies that the wood has been cut from trees within a sustainable forest. For example:

- FSC (Forest Stewardship Council)
- CSA (Canadian Standards Association)
- PEFC (Programme for the Endorsement of Forest Certification)
- MTCC (Malaysian Timber Certification Council)

Training needs for students have also seen major changes, with a much greater emphasis towards work based on the job training – qualifications now available to the Carpenter & Joiner now include:

Qualification	Level Available	Awarding Body
NVQ (National Vocational Qualification)	Wood Occupations: Levels 1, 2, & 3	
SVQ (Scottish Vocational Qualification)	Wood Occupations: Levels 1, 2, & 3	CITB (Construction Industry Training Board) and C&G (City & Guilds)
Construction Award	Wood Occupations: • Foundation • Intermediate • Advanced	

(Continued)

Qualification	Level Available	Awarding Body
IOC Carpentry and Joinery	Carpentry & Joinery: • Foundation • Intermediate • Advanced Craft (MIOC) • Fellow (FIOC)	Institute of Carpenters
Master Certificate Scheme	• Carpentry • Joinery • Shopfitting	Institute of Carpenters The Worshipful Company of Carpenters City and Guilds The Worshipful Company of Joiners and Ceilers

With this revised new edition, we have taken into account all the changes we thought necessary to furnish the reader with current updated information. Together with Volumes 1 & 2 this should meet the requirements necessary for their endeavour to gain enough knowledge within their chosen Wood Occupation (Carpentry and Joinery) to be successful in their examinations and chosen career.

Because of the extensive cross-reference index in the book you should find navigating through it and across the series quite a simple task.

Furthermore, references to Building Regulations, British Standards etc., are included at the end of each section.

This book has been revised to compliment new editions of *Carpentry & Joinery 1* and *Carpentry & Joinery 2*. As with all the books in the series, you will find that due to its highly illustrative content, it is very easy to follow. We believe that a textbook of this nature, deserves well-detailed pictures and annotated diagrams which speak louder than words.

The first edition of this book came into print over 20 years ago and has proved itself time and time again as a valuable source of reference to students of all the wood trades across the world.

Finally, we both feel and ardently believe that Brian's final sentence of the Preface in his first edition to Volume 2, is as important as ever – if not more so now . . . *"Perhaps it could be said that the good craft worker is one who signs his or her work with pride."*

Brian Porter & Christopher Tooke

Acknowledgements

This new Third Edition of *Carpentry and Joinery* would not have been possible without the kind help and assistance of the following companies and organisations who supplied us with technical information and permission to reproduce their artwork and photographs. For this we are both very grateful.

Celuform Ltd, Glass and Glazing Federation, Pilkington (UK) Ltd, Royde and Tucker (Architectural Ironmongery) Ltd, Scandia-Hus Ltd, and Timber Research and Development Association (TRADA).

Company	Figure Number(s)
"balanceuk" Ltd	3.24
Building Research Establishment (BRE)	1.8(a), 1.8(b), 1.8(d), 1.8(e)
Bosch UK Ltd	8.43
Chasmood Ltd	4.89, 4.90
Christopher Addison-British Woodworking Federation (BWF)	4.37, 4.38
Chubb Ltd	4.76(b), 4.96
Clico (Sheffield) Tooling Ltd	8.8
Fisher Fixing Systems Ltd	3.34
Freidland Ltd (Bell push)	4.97
Henderson Garage Doors Ltd	5.13
Ingersoll-Rand Business Ltd (Briton Door Closers)	4.87
Jeld-Wen UK Ltd	3.30, 3.31
Josiah Parks (Union)	4.91, 4.95, 4.96, 4.97
Makintosh and Partners Group Ltd	8.5
National House Building Council (NHBC)	3.43, 3.44
Naco Windows (Ruskin Air Management) Ltd	3.27, 3.28
Reynolds (UK) Ltd (Crompton Hardware, ERA and Yale)	3.65, 4.71, 4.72, 4.79, 4.80, 4.81, 83(b), 4.85, 4.96, 4.97, 5.3 to 5.11, 6.9 to 6.13
Selectaglaze Ltd	3.52
Slik Sliding Door Gear Ltd	10.27, 10.28
Solatone (Door bell)	4.97
Stewart Milne Timber Systems Ltd	1.8(f)

The Bath Knob Shop (tBKS) Architectural Ironmongery	3.61, 3.62, 3.66, 3.67, 4.69, 4.78, 4.88, 4.93
Titon Ltd	3.60, 3.63, 3.64
Trend Machinery & Cutting Tools Ltd	4.62, 7.15, 10.32, 10.33, 10.34, 10.35
Velux Company Ltd	3.36, 3.39, 3.40

Finally, we would both like to thank Hilary Yvonne Porter once again for her continual help, patience and support during the writing of this new third edition.

Prefabricated buildings

In many situations, prefabricated timber buildings provide an alternative to traditional (masonry) on-site building.

The components which go towards constructing load-bearing walls, and in some cases the floors and roof, are either fully or partially prefabricated. This process can be carried out on site, but in the UK it will almost certainly be done within a factory, where conditions can be controlled and operatives are not hindered by unreliable weather.

Once the on-site foundations have been constructed, the pre-made units or components can be delivered and erected as and when required.

The main advantage of prefabrication compared with fabrication on site is that the superstructure can be erected very quickly to form a weather-resistant envelope, thereby providing a sheltered environment for greater work continuity across the building trades.

Whether the building is to be demountable (capable of being dismantled and reassembled elsewhere) or a permanent fixture will mainly depend on its use and size.

1.1 Small demountable timber-framed structures

Structures of this type may be used as site huts for temporary accommodation. Although more commonly used these days, prefabricated, proprietary modular units are generally employed. These not only provide temporary accommodation for site personnel and their equipment, but also offer storage space for perishable and/or valuable building materials.

However, garden sheds and summer-houses are usually constructed as timber-frame structures.

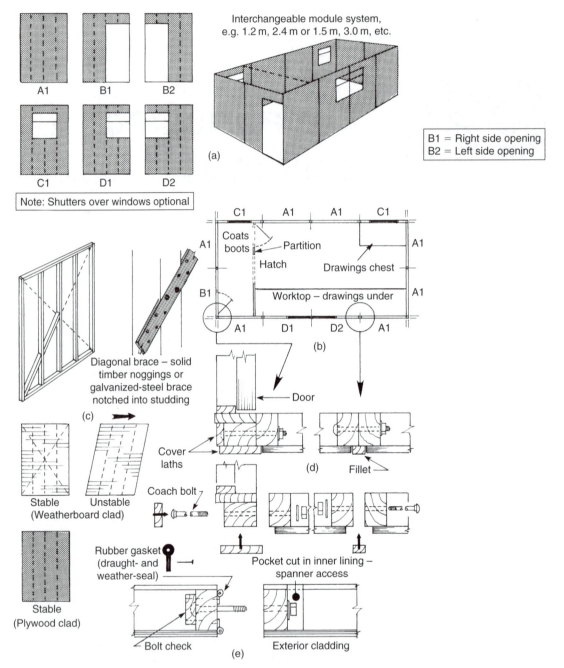

Fig. 1.1 *Timber-framed wall panels.*

The floor, sides, and roof may consist of one or more panels. These panels are joined together at their corners and along their length with coach bolts or similar devices capable of cramping and holding the joint temporarily or permanently secure.

It is common practice to make panels to a modular size (say multiples of 500 mm or 600 mm) to suit board or timber lengths – panels can then be interchangeable as shown in Fig. 1.1a. In this way (provided that structural stability is not affected), a door or window can be sited to suit almost any situation.

Figure 1.1b shows a possible floor layout of a site hut using six panel variations. Except for the partition, the order for the wall panels might read as follows in Table 1.1.

1.1.1 Construction and erection

1.1.1.1 *Walls*

Wall panel sizes vary, being often influenced by:

(a) the type of building;
(b) the size and type of cladding;
(c) transportation;
(d) handling techniques, for example, by hand or machine.

The structural strength of a panel is derived from the row of vertical members called 'studs'. Studding is trimmed at the top by a head-piece, and at the bottom with a sole-piece of the same sectional size. All the joints are nailed.

If panels are to be externally clad with a strip material, e.g. weather-board, etc. (see Chapter 2), the framework should be diagonally cross-braced to prevent 'racking' (lateral movement) as shown in Fig. 1.1c.

Diagonal noggings of solid timber or perforated galvanized-steel strapping, let into and nailed on to studding under the cladding, can be used as bracing as shown in Fig. 1.1c. Diagonal cross-bracing is not always necessary when the frames are clad with a sheet material – an exterior cladding of exterior-grade plywood can provide very good resistance to racking.

Figure 1.1d shows how panels can be joined together. If the panels are lined on the inside – possibly to house some form of insulation material between the studding – pockets should be left to allow access to coach bolts, etc. A more effective joint can be made by introducing a rubber-type gasket between the edges of the panels as shown in Fig. 1.1e – as the bolts are tightened, an airtight and watertight seal is formed.

1.1.1.2 *Floors*

Floor panels should be as large as practicable and decked with tongued-and-grooved boards or exterior-grade plywood. Panel sizes

Table 1.1   Order of wall panels
Site hut (see Fig. 1.1) – wall panels 1.2 m (type 600 module)

Panel code	Number
A1	7
B1	1
B2	0
C1	2
D1	1
D1	1
Total number of panels	12

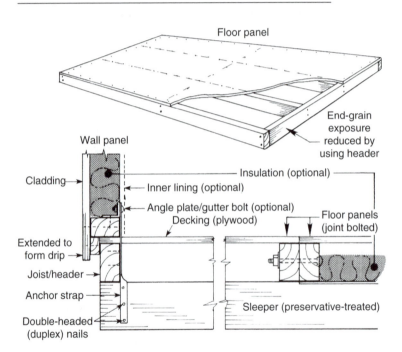

Fig. 1.2   *Floor panels and wall-to-floor details*

must be adaptable to wall modules, and provision must be made for bolting panels together and for anchoring the walls to them, as well as for fastening down to ground sleepers, as shown in Fig. 1.2. Floor panels may also incorporate thermal insulation (see also Volume 2, section 5.3 Thermal insulation).

1.1.1.3 Roofs

Whole roof sections can be prefabricated as shown in Fig. 1.3a to drop on to and be anchored to wall sections via bolt-on angle plates or steel straps. Joists may be cut from the solid, laminated or, for lightness, fabricated using the box–beam method (a solid timber framework clad both sides with an exterior-grade plywood). Thermal insulation may be incorporated within roof sections (but see Volume 2,

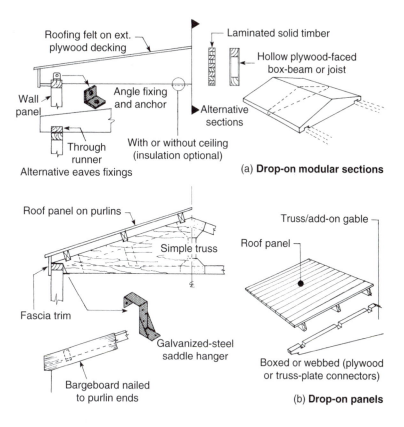

Fig. 1.3 *Roof details (Note: thermal insulation not shown)*

section 8.10, with regard to avoiding interstitial condensation, i.e. condensation occurring within the fabric of the building).

Alternatively, roof panels can be supported by simple trusses evenly spaced between add-on gable ends. As shown in Fig. 1.3b, trusses can be built up as a box section or as a timber framework with joints webbed with exterior-grade plywood or steel truss-plate connectors. Trusses can be set and anchored to wall panels, or suspended from them by anchored steel saddle-hangers.

Gable-end sections support the panel ends. Notches are cut to receive runners (purlins), their ends being covered (trimmed) with a bargeboard.

1.1.1.4 Sequence of erection (see Fig. 1.4)

1. Lay and level heavy-sectioned sleepers (bearers). Wood-preservative treatment is essential. (See also Volume 1, Chapter 3.)
2. Lay and connect (using coach bolts) an appropriate number of floor panels. Anchor them to the sleepers.
3. Erect and bolt together two corner panels. Fix them to the floor and temporarily brace their top corner.

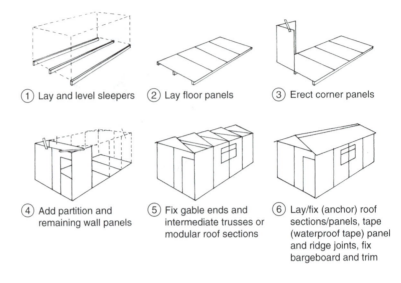

① Lay and level sleepers   ② Lay floor panels   ③ Erect corner panels

④ Add partition and   ⑤ Fix gable ends and   ⑥ Lay/fix (anchor) roof
  remaining wall panels   intermediate trusses or   sections/panels, tape
                              modular roof sections   (waterproof tape) panel
                                                          and ridge joints, fix
                                                          bargeboard and trim

Fig. 1.4 *Site hut assembly details*

4. (a) Form the adjacent end corner;
   (b) Add the internal partitions (if any);
   (c) Loosely bolt on the remaining panels;
   (d) Check for alignment and squareness – add temporary corner braces;
   (e) Tighten all nuts. Fix to the floor.
5. (a) Bolt on gable sections and fix intermediate trusses; or
   (b) Drops on and bolt together whole roof sections.
6. (a) Lay and fix roof panels;
   (b) Tape over the panel and ridge joints with waterproof self-adhesive tape;
   (c) Fix trim – the eaves and barge-board.

1.2 Timber-framed dwellings

These are permanent buildings consisting of an outer timber framework structurally designed to support an upper floor and/or a roof structure. All loads are transmitted to the foundations via the timber-studded framework. This studded wall framework can be built up by one of two methods:

(i) balloon-frame construction, or
(ii) platform-frame construction.

In balloon-frame construction (Fig. 1.5), which is generally not used much in the UK, the wall panels extend in height from the ground floor to the eaves. The upper floor is suspended from within the full-height studded wall structure. Load-bearing partitions can offer mid-support.

In platform-frame construction (Figs 1.6 and 1.7) storey-height wall panels are erected off the ground-floor platform. These load-bearing panels provide support for the upper-floor platform, off which the upper wall panels are erected.

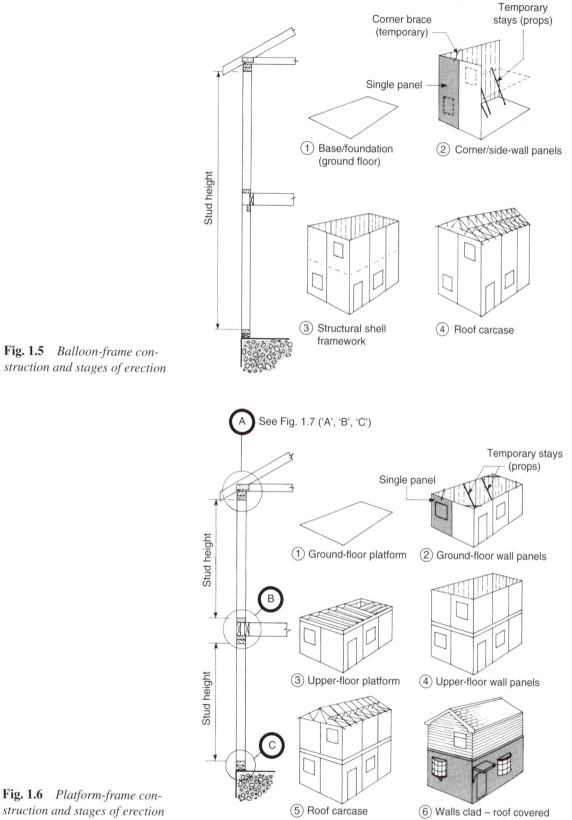

Fig. 1.5 *Balloon-frame construction and stages of erection*

Corner brace
(temporary)

Temporary
stays (props)

Single panel

① Base/foundation
(ground floor)

② Corner/side-wall panels

③ Structural shell
framework

④ Roof carcase

Ⓐ See Fig. 1.7 ('A', 'B', 'C')

Stud height

Stud height

Ⓑ

Ⓒ

Temporary stays
(props)

Single panel

① Ground-floor platform

② Ground-floor wall panels

③ Upper-floor platform

④ Upper-floor wall panels

⑤ Roof carcase

⑥ Walls clad – roof covered

Fig. 1.6 *Platform-frame construction and stages of erection*

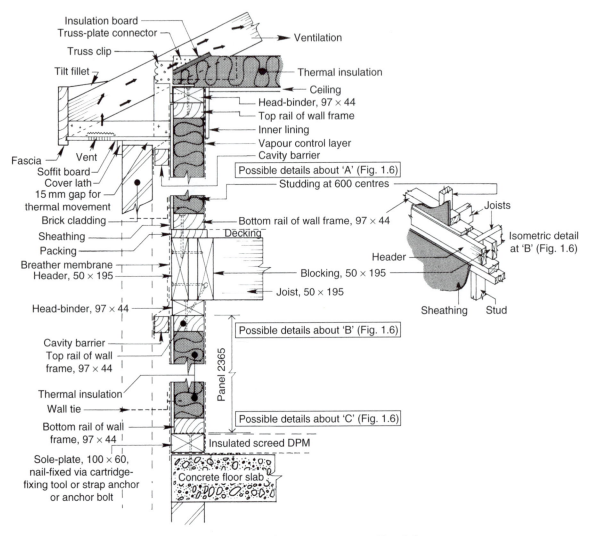

Fig. 1.7   *Vertical section through platform-framed construction – see Fig. 1.6*

Figure 1.8a–f illustrates the main stages in the erection of a partially off-site prefabricated timber-frame house. (Not all the photographs are of the same house.) It should be noted that in these cases the breather membrane (as shown in Fig. 1.8(f)) is fixed over the face of the panels *in-situ*; in this way the probability of a pre-covered panel becoming damaged while in transit or stored on site is avoided. However, many manufacturers prefer to pre-cover their panels, to ensure early weather protection to the building as a whole.

With both methods, the roof structure bears upon load-bearing wall panels.

Nearly all timber-framed houses in the UK are built using the platform-frame construction method; therefore the text that follows refers to this method.

(a) Soleplate in place, used to provide a floor screed. Notice the provision for service pipes

(b) Exterior corner panels being erected

(c) Interior studwork showing temporary bracing for stability

(d) Roof trusses pre-assembled over ground floor slab – prior to being lifted up in place

(e) Part of pre-assembled roof being lifted and lowered into position

(f) Exterior cladding in progress on left hand construction

Fig. 1.8 *Erection of a timber-framed house*

1.2.1 Wall construction

Panels may be constructed by one of three methods:

(i) *Stick built*: this is the least popular method, which involves fabricating all the panels on site.

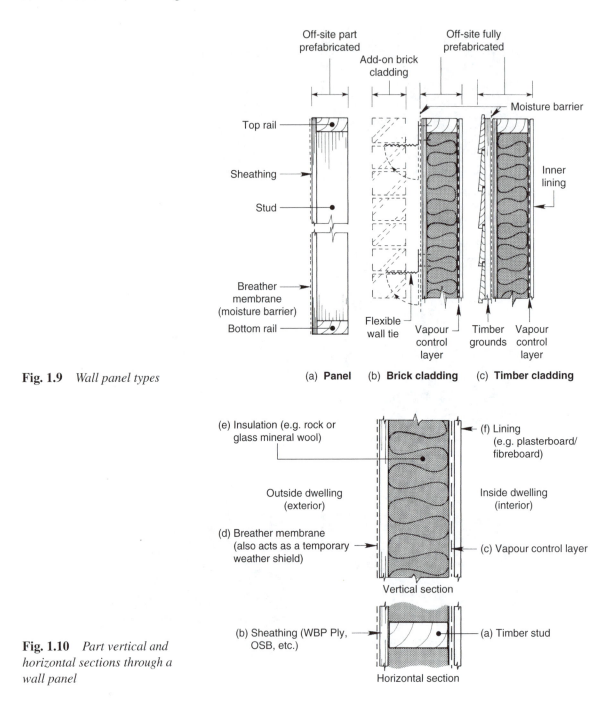

Fig. 1.9   *Wall panel types*

(a) **Panel**   (b) **Brick cladding**   (c) **Timber cladding**

Fig. 1.10   *Part vertical and horizontal sections through a wall panel*

(ii) *Partial off-site prefabrication* (Fig. 1.9a): a jig-built studded framework is clad with a sheathing (sheet material – usually exterior-grade plywood see Volume 1, section 4.1.2) to prevent racking and is then often covered with a breather membrane (see Figs 1.10 and 1.11). Insulation, a vapour control layer, and

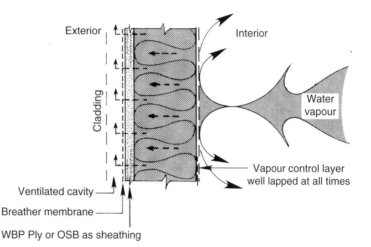

Exterior

Interior

Cladding

Water vapour

Ventilated cavity

Vapour control layer well lapped at all times

Breather membrane

Fig. 1.11 *Providing a vapour check*

WBP Ply or OSB as sheathing

an inner lining are fixed on site after the house has been made watertight. The panel size will depend on the means of transport and the site-hoisting facilities available.

(iii) *Fully off-site prefabricated panels*: if the building is to be clad with masonry (i.e. brickwork), then the panels will include wall ties prefixed to the studs via breather-membrane-covered sheathing, bitumen-impregnated fibreboard, or OSBs (orientated strand board) (see Volume 1, Table 4.10), as well as insulation, a vapour control layer, and the inner lining, as shown in Fig. 1.9b.

Timber-clad panels (Fig. 1.9c) could include pre-fixed windows.

Figure 1.10 shows part vertical and horizontal sections through a completed panel. Components include:

(a) *Structural timber studs*: these are usually 'structural'-graded softwood, sawn or (more likely) planed finish. Preservative treatment will be required in most cases. Hardwoods may also be used. The stud sections are 38 mm minimum finished thickness (usually 45 mm), with a width of 95 mm finish (or wider, to accommodate the necessary thickness of thermal insulation material).

(b) *Sheathing*: the most common sheet materials are:
  (i) *Plywood* – structural sheathing quality, bonded to WBP standard (see Volume 1, section 4.1.2).
  (ii) *OSB (orientated strand board)* – multi-layer board, cross-banded in a similar fashion to 3-ply-veneer plywood (see Volume 1, section 4.5).
  (iii) *Medium board* – should be stiff enough to provide racking resistance (see Volume 1, section 4.6.3).
  (iv) *Bitumen-impregnated fibreboard* – racking resistance of the panel must be provided by other means (see Volume 1, Table 4.10). Provided that the panel joints are sealed, breather membrane may be omitted.

(c) *Vapour control layer*: this is usually a 250 or 500 gauge polythene membrane fixed to the warm side of the insulation (see also Volume 2, Fig. 5.8). All joints should meet on studs or noggings and should overlap by at least 100 mm. Stainless steel staples should be used.

(d) *'Breather' membrane*: this is a waterproof membrane which helps keep the timber-framed shell weathertight but, because of its permeable nature, allows water vapour to escape from inside the dwelling as and when required to do so.

(e) *Insulation material*: this must be a permeable type of either rock wool or glass mineral wool. A thickness of 90 mm would give a U-value (see below) of about 0.32 W/(m^2°K) through the whole wall structure (including cladding), which is nearly double the statutory requirement to satisfy the Building Regulations Approved Document L1, 2002.

(f) *Lining*: suitable lining materials and the requirements regarding surface spread of flame are the same as those for partitions, as discussed in Volume 2, section 9.4.2.

> Note: *Temperature difference may also be expressed in celsius (C). As a temperature difference of 1 C = 1 °K, a U-value is the same whether expressed in W/(m^2°C) or W/(m^2 K).*

Figure 1.11 shows how the vapour check is made – its presence should prevent moisture in the warm air entering the insulation-filled (or partly filled) cavity and condensing out as it reaches the dew point (the temperature at which dew begins to form) at or near the cold outer surface. An inadequate vapour control layer could result in the insulation becoming wet and ineffective and increase the moisture content of the timber framework. However, because it is almost impossible to achieve a 100% vapour check, those small amounts of vapour which do pass through should be allowed to continue unhindered via the permeable breather membrane into a ventilated cavity (all forms of cladding should be vented behind) where they can disperse freely.

Figure 1.12a shows a skeleton view of a panel, with a lintel over a window opening, being connected to a similar panel at a corner. All panel joints are butted together – the method of securing the joints will depend on the panel type and corner details, etc. (Figs 1.12b and c). Further to nailing and using metal corner straps, the timber head-binders should lap all the panel joints, both at corners and on the straight.

In Fig. 1.12c it can be seen how provision has been made for services (electric and telephone cables, etc. in suitable trunking) by rebating the edges of each panel. A plough groove houses a loose tongue which serves as a location aid, panel stiffener, and draught-seal. Rubber draught-seals can be provided at corners and floor levels.

Panels can be secured at ground-floor level via a timber sole-plate previously fixed to the concrete floor slab (see Fig. 1.7). (Fixing to a timber-suspended ground floor is shown in Volume 2, Fig. 5.8.) Upper floor and roof connection details are also shown in Fig. 1.7. (See also Volume 2, Fig. 7.6, for arrangements with a flat roof.)

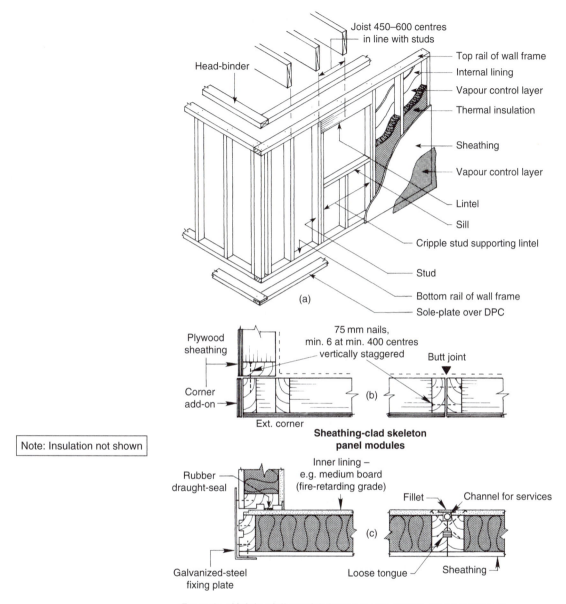

Note: Insulation not shown

Fig. 1.12 *Timber-framed wall connection details*

1.2.2 U-value

U-value, or 'thermal transmittance', provides a measure of the rate at which heat will flow through a building structure between two environments at different temperatures. It is the rate of heat flow (in watts) through one square metre of the structure when the environmental temperatures on each side differ by 1°K, and it therefore has the units watts per metre squared per °C or $W/(m^2\,°K)$.

Each component in the structure, and the air layer on each side of the structure, has its own thermal resistance to heat flow (see below), and the U-value of the assembled structure is calculated from

$$\text{U-value} = \frac{1}{R_{\text{total}}}$$

where R_{total} is the sum of all the thermal resistances offered by the components of the structure, and the air which surrounds them, at any given point in the structure. To find the thermal resistance of a material, we must know its thermal conductivity and hence its thermal resistivity.

1.2.3 Thermal conductivity (*k*)

This gives an indication of an individual material's ability to conduct heat. It is the rate of heat flow (in watts) through one square metre of a sample of material one metre thick when there is a temperature difference of 1°K between the surfaces of the material; it therefore has

Table 1.2 Typical values of thermal conductivity and thermal resistivity for various materials and products

Material	Thermal conductivity (*k*) (W/(m °K))	Thermal resistivity (1/*k*) (m °K/W)
Air	0.026	38.46
Expanded polystyrene	0.034	29.41
Balsa wood	0.040	25.00
Fibreglass (glass-wool fibre)	0.040	25.00
Fibreboard – insulation and medium board (varies with density and m.c.)	0.053–0.065	15.38–18.87
Bitumen-impregnated fibreboard	0.070	14.29
Hardboard (fibreboard)	0.080	12.50
Aerated-concrete building blocks (varies with density)	0.110–0.196	5.10–9.09
Softwood	0.130	7.69
Hardwood	0.150	6.67
Chipboard	0.150	6.67
Plasterboard	0.160	6.25
Plastering	0.380	2.63
Glass	1.050	0.95
Brick work – commons (varies with density)	1.250	0.80
Concrete (dense)	1.430	0.70
Steel	50.00	0.02
Aluminium alloy	160.00	0.006

the units $(W/m^2)/(m\,°K)$ or $W/(m\,°K)$, i.e. watts per metre per $°K$ (or $W/(m\,C)$ if temperature differences are quoted as celsius).

Manufacturers of building materials quote the thermal conductivity as an indication of a material's thermal performance – the lower the thermal conductivity, the greater the material's resistance to heat flow through it. For example, typical thermal conductivities might be:

for brickwork $1.25\,W/(m\,°K)$
for woodwork (solid) $0.13\,W/(m\,°K)$

Therefore wood has a much greater resistance to heat transfer, i.e. it is a better insulator, than brickwork. Table 1.2 gives other typical values of thermal conductivity.

1.2.4 Thermal resistivity (r)

This gives an indication of a material's ability to resist the flow of heat and is defined as the reciprocal of thermal conductivity, i.e. $1/k$. Its units are therefore $m\,°K/W$ (or $m\,C/W$).

1.2.5 Thermal resistance (R)

For a particular thickness of material, this gives the resistance to heat flow through one square metre of the material when there is a temperature difference of $1\,°K$ between the surfaces of the material. It has the units $m^2\,°K/W$ (or $m = C/W$) and is calculated by multiplying the thermal resistivity by the thickness of the material. For example, for glass fibre with a thermal conductivity (k) of $0.04\,W/(m\,°K)$,

$$\text{thermal resistivity} = \frac{1}{k}$$

$$= \frac{1}{0.04\ W/(m°K)} = 25\ m°K/W$$

For a thickness of 90 mm (0.09 m) of this glass fibre,

$$\text{thermal resistance} = \text{thermal resistivity} \times \text{thickness}$$
$$= 25\ m\,°K/W \times 0.09\ m = 2.25\ m^2\,°K/W$$

Values of thermal resistance can be used to calculate the U-value of a structure. It is usual to take arbitrary figures for the thermal resistances of the air layers on each side of the structure (remember that U-values relate to differences between environmental temperatures, not surface temperatures) and for any air gaps within the structure, as shown in the example below.

Assuming that the section through the wall panel shown in Fig. 1.10 consists of the following (disregarding the vapour control layer and the breather membrane):

Material	Thickness
Plasterboard	12.5 mm
Glass-fibre insulation (mineral wool)	90 mm
Plywood sheathing	9.5 mm

then the U-value of the panel can be determined by first calculating the thermal resistance of each element in the structure, using the values for k given in Table 1.2.

For example, for the plasterboard

$$k = 0.16 \, \text{W/(m°K)} \quad \text{from Table 1.1}$$

$$\therefore \quad \frac{1}{k} = \frac{1}{0.16 \, \text{W/(m°K)}} = 6.25 \, \text{m°K/W}$$

$$\text{thickness} = 1.27 \, \text{mm} = 0.0127 \, \text{m}$$

$$\therefore \quad R = \frac{1}{k} \times \text{thickness}$$

$$= 6.25 \, \text{m°K/W} \times 0.0127 \, \text{m}$$

$$= 0.08 \, \text{m}^2\text{°K/W}$$

The following table can be drawn up:

Material	Thick-ness, T (m)	Thermal conduc-tivity, k (W/(m°K))	Thermal resis-tivity, $1/k$ (m°K/W)	Thermal resistance, $R = T \times 1/k$ (m^2°K/W)
Plasterboard	0.0127	0.16	6.25	0.08
Glass-fibre insulation	0.09	0.04	25	2.25
Air space within cavity (arbitrary value)				0.18
Plywood sheathing	0.0095	0.14	7.14	0.07
Inner and outer surface air layers (arbitrary value)				0.18
Total thermal resistance				2.76

Note: *In the above calculation, no allowance has been made for sections including a stud. If external cladding had also been included, the overall U-value would have been in the region quoted earlier (0.32 W/(m² °K)).*

For the panel section, therefore

$$U\text{-value} = \frac{1}{R_{\text{total}}}$$
$$= \frac{1}{2.76\,\text{m}^2\,°\text{K/W}}$$
$$\approx 0.36\,\text{W/(m}^2\,°\text{K)}$$

1.2.6 Cavity barriers (see Fig. 1.7)

These provide a means of closing the cavity and limiting the volume of the cavity between the timber framework and any external cladding (brickwork, etc.) or connecting elements – in the event of a fire, the cavity could otherwise allow smoke and flame to spread through or around the structure. Location and distribution of cavity barriers must satisfy the requirements of the Building Regulations Approved Document B, Fire Safety: 2000.

A cavity barrier may, for example, be constructed of:

(a) timber not less than 38 mm thick;
(b) steel not less than 5 mm thick;
(c) calcium silicate, cement-based or gypsum-based plasterboard not less than 12.0 mm thick;
(d) wire-reinforced mineral wool blanket not less than 50 mm thick.

1.2.7 Fire stops

These are seals of non-combustible material used to close an imperfection of fit between building elements and components, for example, where smoke and flame could otherwise pass. For specific locations, see the Building Regulations Approved Document B, Fire Safety: 2000.

1.2.8 Fire spread

In the event of a possible fire, timber used as an external cladding may increase the likelihood of the fire spreading to other buildings. Generally, domestic buildings must be separated by a space of at least 1 m and restricted to wall areas of 3 storeys in height and 24 m in length. See the Building Regulations Approved Document B, Fire Safety: 2000.

1.2.9 Volumetric house units (whole or part off-site house prefabrication)

Over the past few years there has been a slight increase in this form of building construction, where the units are made under factory conditions, in a dry environment and quality controlled. This reduces the risk of on-site defects associated with traditional forms of construction.

Once prefabricated in the factory, these large pre-finished sectional units are then connected together on site to form a whole house – a highly specialized operation requiring special production techniques, transportation, and lifting facilities.

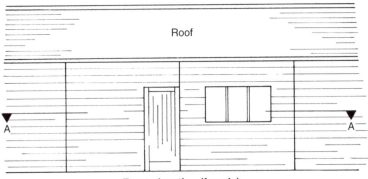

Roof

Front elevation (façade)

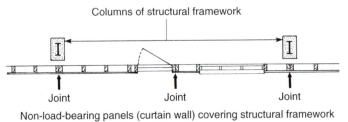

Columns of structural framework

Joint   Joint   Joint

Fig. 1.13   *Timber-framed panels forming a curtain wall*

Non-load-bearing panels (curtain wall) covering structural framework

Horizontal section A-A

Note: *With this method of building, the main load-bearing structure must be provided with lateral support and resistance to racking.*

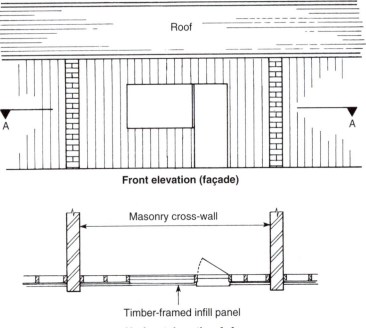

Roof

Front elevation (façade)

Masonry cross-wall

Timber-framed infill panel

Fig. 1.14   *Timber-framed infill panels*

Horizontal section A-A

1.2.10 Standard Assessment Procedures (SAPs)

Originally introduced in 2001 (as part of Approved Document L1: Conservation of Fuel and Power) with the latest revision in 2006, its purpose is to determine the energy efficiency on all new construction in relation to insulation, heating, lighting, etc.

A SAP rating may vary between 1 and 100, indicating the higher the range the more energy efficient the property.

Furthermore, it will give the prospective home owner an idea of the probable running cost of the property and its effects on the environment.

1.3 Timber-framed curtain walls

These form a non-structural façade to cover a framed load-bearing structure. An example of how light timber-framed panels can be used for this purpose is shown in Fig. 1.13. Although the panels are lighter, their construction and insulation techniques are similar to those already mentioned. The panels will rely on the main substructure for their fixing and support.

1.4 Timber-framed infill panels

By using masonry cross-walls (either end or party walls) to support all the main load-bearing parts of a building, for example the floors and the roof, the front and back walls can, in some situations, be made as an infill panel (or panels) as shown in Fig. 1.14.

References

Building Regulations Approved Document L1, Conservation of Fuel and Power in Dwellings, 2002 (SAPs).
Building Regulations Approved Document B, Fire safety: 2000.
BS 5268–6.1:1996, Structural use of timber. Code of practice for timber walls. Dwellings not exceeding four storeys.
BS 1282: 1999, Wood Preservative. Guidance on choice use and application.
BS 4072: 1999, Preparations for wood preservatives.
BS 4261: 1999, Wood Preservation. Vocabulary.
BS 5589: 1989, Code of practice for wood preservation.
BS 5268–5: 1989, Structural use of timber, code of practice for the preservative treatment of structural timber.
BS 5707: 1997, Specification for preparations of wood preservatives in organic solvents.
TRADA Timber-framed construction.
BRE Timber-framed construction information sheet.
NORDIC Timber construction buildings.
BBA (British Board of Agreement).

2 External cladding

No matter what form it takes, cladding provides a non-load-bearing weather protection and in most cases a decorative finish to whatever structure it is required to cover.

2.1 Cladding properties

Any material may be used as a cladding provided that it is available in an acceptable form, and that it possesses the necessary properties, for example:

(a) weather resistance;
(b) durability;
(c) strength to resist impact;
(d) resilience to abrasion;
(e) fire performance;
(f) thermal insulation;
(g) minimal maintenance

and that it complies with current regulations (Approved Document B4 2000, Fire spread).

The relationship between the material, its form and type, is shown in Table 2.1 with reference to Fig. 2.1.

In the UK, brickwork is more often than not the main external cladding used with timber-framed houses. However, if other cladding materials are not used all over, they are often included in the design and used alongside brickwork or stonework as a form of visual relief. Examples of how this could be achieved are shown in Fig. 2.2.

2.2 Timber cladding

Timber used as a cladding material has been common practice since the early days of house building as a protection against the elements. Depending on the timber species, the climatic conditions, and any preservative treatment, its life span can exceed 50 years – any boards that do show signs of decay or defect at a later date can easily be replaced.

Table 2.1 Cladding materials, their form/type and application

Cladding material	Form/type	Application	Figure reference
Softwood timbers which have been preservative treated, plus some varieties of hardwood (see Volume 1, Tables 1.14 and 1.15)	Square edged Matchboard Profiled weatherboard	Vertically Vertically Horizontally Diagonally (chevron)	2.1 (a) 2.1 (a) 2.1 (b) 2.1 (c)
Aluminium	Profiled strips	Vertically	2.1 (a)
Steel (treated or coated)	Strip and full sheets (profiled)	Vertically Horizontally Diagonally	2.1 (a) 2.1 (b) 2.1 (c)
Plastics (PVC-u) (see Fig. 2.15)	Profiled weatherboard	Vertically Horizontally Diagonally	2.1 (a) 2.1 (b) 2.1 (c)
MDF (external grade, see Volume 1, section 4.6.6)	Matchboard Profiled weatherboard Full sheet	Vertically Horizontally Diagonally (chevron) Panel	2.1 (a) 2.1 (b) 2.1 (c) 2.1 (d)
Non-combustible board (magnesium-based cements) (see Volume 1, section 4.8)	Strips Full sheets	Horizontally Panel	2.1 (b) 2.1 (d)
WBP Plywood (see Volume 1, Chapter 4, section 1.2)	Strips Full sheet	Horizontally Panel	2.1 (b) 2.1 (d)
Tempered hardboard (see Volume 1, section 4.6.5)	Strips Full sheet	Horizontally Panel	2.1 (b) 2.1 (d)
Western red cedar (softwood) (preservative treated or untreated heartwood) (see Volume 1, Table 1.14)	Profiled weatherboard or Hung shingles Matchboard Board or Board/Batten	Horizontally Vertically hung Vertically Diagonally (chevron) Vertically	2.1 (b) 2.1 (e) 2.1 (a) 2.1 (c) 2.1 (a)
Cement-bonded particle board (see Volume 1, 4.4)	Strips and full sheets	Panel	2.1 (e)
Clay	Hung tiles	Vertically hung	2.1 (e)
Concrete	Hung tiles (square edged, and patterned)	Vertically hung	2.1 (e)
Slate (natural and synthetic)	Hung slates	Vertically hung	2.1 (e)
Cement rendering on steel mesh	Rendering (flat, textured pattern surface)	Built-up surface covering	2.1 (f)
Masonry	Brickwork, blockwork and stonework	Built-up surface covering	2.1 (f)

Note: Reference should be made to Fig. 2.1, which shows the format in which the above materials appear. Also see Figs 2.3 and 2.4.

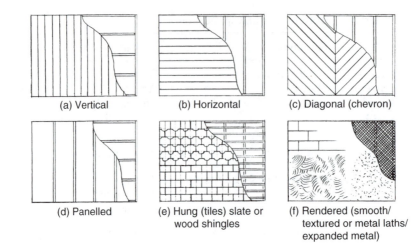

(a) Vertical   (b) Horizontal   (c) Diagonal (chevron)

(d) Panelled   (e) Hung (tiles) slate or wood shingles   (f) Rendered (smooth/ textured or metal laths/ expanded metal)

Fig. 2.1 *Cladding formats (see Table 2.1)*

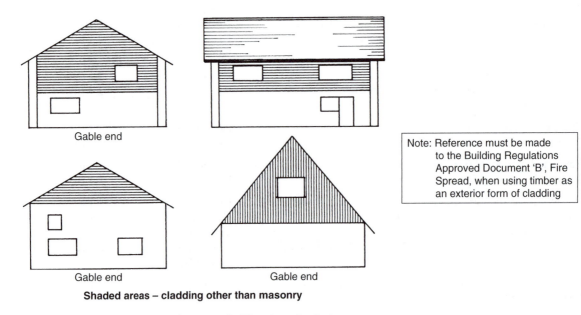

Gable end

Gable end   Gable end

Note: Reference must be made to the Building Regulations Approved Document 'B', Fire Spread, when using timber as an exterior form of cladding

Shaded areas – cladding other than masonry

Fig. 2.2 *Incorporating more than one cladding into the design*

2.2.1 Fixing battens (grounds)

Timber cladding in most cases is fixed to horizontal or vertical timber battens (grounds), which are preservative treated, with any cut ends also treated before fixing.

A common size for grounds is 38 × 25 mm, which is fixed to the studwork, in the case of timber-framed construction, through the breather membrane and spaced at a maximum of 600 mm (see Fig. 2.5). By using grounds, it also provides a cavity to assist with ventilation (see section 2.2.4).

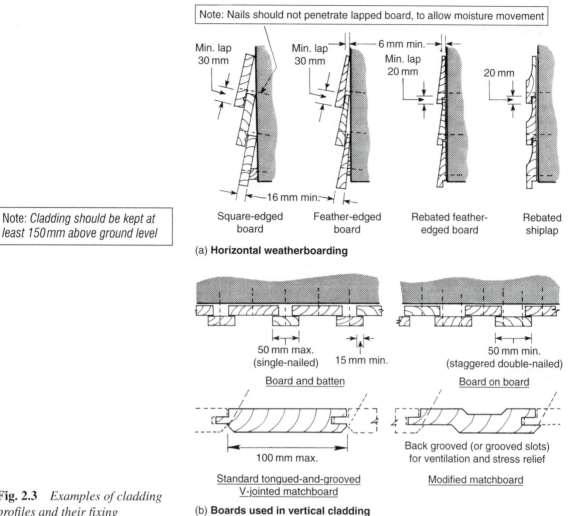

Note: Nails should not penetrate lapped board, to allow moisture movement

Min. lap 30 mm

Min. lap 30 mm

6 mm min.

Min. lap 20 mm

20 mm

16 mm min.

Square-edged board

Feather-edged board

Rebated feather-edged board

Rebated shiplap

(a) **Horizontal weatherboarding**

Note: *Cladding should be kept at least 150 mm above ground level*

50 mm max. (single-nailed)

15 mm min.

50 mm min. (staggered double-nailed)

Board and batten

Board on board

100 mm max.

Back grooved (or grooved slots) for ventilation and stress relief

Standard tongued-and-grooved V-jointed matchboard

Modified matchboard

(b) **Boards used in vertical cladding**

Fig. 2.3 *Examples of cladding profiles and their fixing*

2.2.2 Timber profiles

To limit the possible entry of water between joints and overlaps, cladding profiles in all cases are designed and fixed in such a way that any water is quickly dispersed. A variety of common profiles is available from most timber merchants with sizes ranging from 22 mm thick for tongue-and-groove boarding (matchboarding), and 16 mm for feather-edge boarding reduced to 6 mm on one edge.

Details of square-edged boards together with examples of weatherboarding, which is usually fixed horizontally, are shown in Fig. 2.3a. Matchboarding or square-edged boards are fixed vertically (Fig. 2.3b), or diagonally, to form a chevron pattern (Fig. 2.1c).

Figure 2.3 shows some examples of cladding profiles in relation to their fixings.

2.2.3 Fixing timber cladding

When fixing, allowance must be made to allow the timber to expand and contract during changes in climatic conditions that would otherwise lead to the cladding splitting. For this reason, fixing must not be through the area where the cladding overlaps, see Fig. 2.3a.

Timber cladding may be fixed to the structure in three ways:

   (i) *Horizontally* (Fig. 2.3a): visually appears to lengthen yet reduce the height of the cladded area. (This is the most popular method used in the UK.).

  (ii) *Vertically* (Fig. 2.3b): produces the reverse effect, making the cladded area appear taller. Vertical fixing is preferred, as it allows water to be shed from the surface more rapidly.

 (iii) *Diagonally* (Fig. 2.1c): produces a pleasing effect; at the same time it can assist water dispersal.

Cladding that has been quarter sawn, offers less movement compared to tangentially sawn timber, particularly in vertical cladding, see Fig. 2.4.

2.2.4 Venting timber cladding

Whether the cladding is fixed to a timber-framed structure (as shown in Fig. 2.5 and Fig. 1.9), or via timber battens (grounds) fixed to masonry (when a breather membrane is usually omitted), it will require back-venting with a gap of at least 25 mm to ensure the safe venting away of any moisture which may enter the cavity, be it from outside the cladding or from the vapour pressure from inside the dwelling. Otherwise both the cladding and the wall it serves could suffer the effect of permanent dampness.

Additionally, any vented areas must include a fire stop (see Fig. 2.6), as required by the Building Regulations, to prevent the possibility of any fire spread between dwellings.

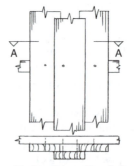

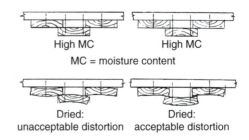

Fig. 2.4 *Moisture movement in vertical cladding*

**Horizontal section A-A
stable quarter sawn boards**

**Horizontal section A-A
tangential sawn boards**

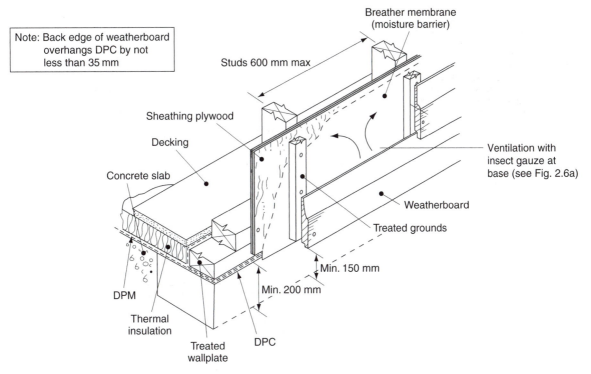

Note: Back edge of weatherboard
 overhangs DPC by not
 less than 35 mm

Breather membrane
(moisture barrier)

Studs 600 mm max

Sheathing plywood

Decking

Concrete slab

Ventilation with
insect gauze at
base (see Fig. 2.6a)

Weatherboard

Treated grounds

Min. 150 mm

DPM

Min. 200 mm

Thermal
insulation

Treated
wallplate

DPC

Fig. 2.5   *Vented weatherboard to a timber-frame house*

Figure 2.6 shows how back-venting can be provided at this level. Provided that gaps are left under window sills and at the eaves, natural venting is easily obtained.

Horizontal cladding (Fig. 2.6a) will require vertical preservative-treated back battens (grounds) – these do not restrict venting (see also section 2.2.1).

Venting is restricted with a vertical cladding of standard matchboard on horizontal battens (Fig. 2.6b); however, this can be overcome by leaving staggered gaps or by cutting notches along each run of back batten, or by using counter battens (crossing over) as shown in Fig. 2.6c. On the other hand, board-and-batten and board-on-board vertical cladding will be self-venting, due to the gaps behind each cover board. As with horizontal cladding, gaps at the top and base of the cladding (eaves and sills, etc.) will still be required.

Figure 2.6 deals in the main with venting, but fixings about ground level are also shown to overcome problems associated with ground vegetation and the incidence of water splashing. For these reasons, a minimum gap of a 150 mm is left between the first board and ground level (see also Fig. 2.5).

Note: If a cavity barrier is required
(e.g. building regulations) at
'X', wired mineral wool can be
positioned as necessary.
This should still allow the back
cavity to breathe. Fly-proof
gauge can be fixed across
all vent openings

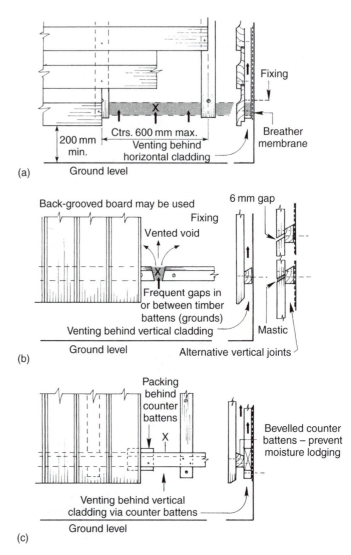

(a)

Fixing

Breather
membrane

200 mm
min.

Ctrs. 600 mm max.

Venting behind
horizontal cladding

Ground level

(b)

Back-grooved board may be used

6 mm gap

Fixing

Vented void

Frequent gaps in
or between timber
battens (grounds)

Venting behind vertical cladding

Mastic

Ground level

Alternative vertical joints

(c)

Packing
behind
counter
battens

X

Bevelled counter
battens – prevent
moisture lodging

Venting behind vertical
cladding via counter battens

Ground level

Fig. 2.6   *Fixing and providing a ventilated cavity*

Note: *By venting the back of the cladding, a vented cavity has been formed which may be regarded by the local authority and the Building Regulations as a fire hazard. If this is the case, a cavity barrier of wired mineral wool (see the note in Fig. 2.6) may be acceptable to restrict the movement of smoke or flame (see section 2.2.6)*

2.2.5 Durability

Nearly all commercial softwood will need treating with a wood preservative (timber derived from the heartwood of Western red cedar is one exception). The heartwoods of many hardwoods may be used in their natural state, without the use of a wood preservative, for example,

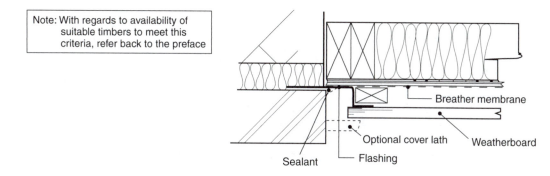

Note: With regards to availability of suitable timbers to meet this criteria, refer back to the preface

Breather membrane

Optional cover lath   Weatherboard

Sealant   Flashing

Fig. 2.7   *Horizontal details at junction between cladding and masonry*

Sapele, European Oak and Utile. (See Volume 1, Table 1.15 for further examples.)

2.2.6 Fire performance

Restrictions are placed on areas of timber cladding in close proximity to a boundary (see Building Regulations Approved Document B, Fire spread: 2000) for fear of fire spreading from another building or other source by way of radiant heat or flying embers, etc. It is really a question of calculating all the 'unprotected areas' (for example window and door openings and cladding), before the boundary distances to satisfy the Building Regulations can be determined for that amount of unprotected area – or vice versa.

2.3 Detailing

Butting horizontal cladding up against masonry should be avoided – a suitable gap should be left as shown in Fig. 2.7, so that end-grain treatment can be maintained with ease.

Corner detailing is important – not only because of appearance but also because of end-grain exposure.

Figure 2.8a shows examples of corner details for vertical board-on-board cladding, with Fig. 2.8b showing corner details for horizontal weatherboard cladding.

Figures 2.9 to 2.11 shows how horizontal and vertical cladding can be finished around a window opening. Extra ventilation under the sill of a wide window can be achieved by cutting a grill into the under-sill batten or by planting a weather-strip (board) on to the batten and widening the abutting gaps.

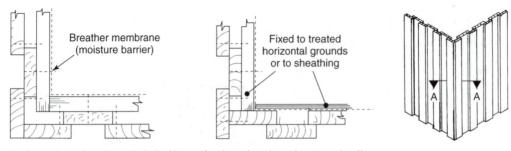

Horizontal section through A-A: Alternative board on board corner details

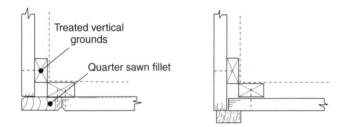

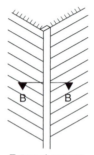

Horizontal section through B-B: Alternative weatherboarding corner details

External corners

Fig. 2.8   *Vertical and horizontal boarding (board-on-board) typical corner details*

2.4 Timber tiles (Figs 2.1e and 2.12)

Timber tiles are generally manufactured from Western Red Cedar (see Volume 1, Table 1.14) and used either for roofing or cladding. Although naturally durable they will have been preservative treated prior to purchasing.

Figure 2.12a shows how they are cut from the log (quarter sawn) to form either:

(i) shingles;
(ii) shakes.

2.4.1 Shingles (Fig. 2.12c)

Shingles are produced by cutting the log to the required length (Fig. 2.12a), usually 400 to 600 mm, and then quarter sawn to produce straight grain – to prevent cupping. The block is then sawn to produce tapered shingles of 75 to 355 mm in width with 2 to 4 mm at their thin edge (Fig. 2.12b). Additionally, shingles may have a profiled (rounded) face known commonally as 'fish scale' faced.

2.4.1.1 Fixing

Shingles are fixed to horizontal grounds (see sections 2.2.1 and 2.6), which are spaced according to length and type.

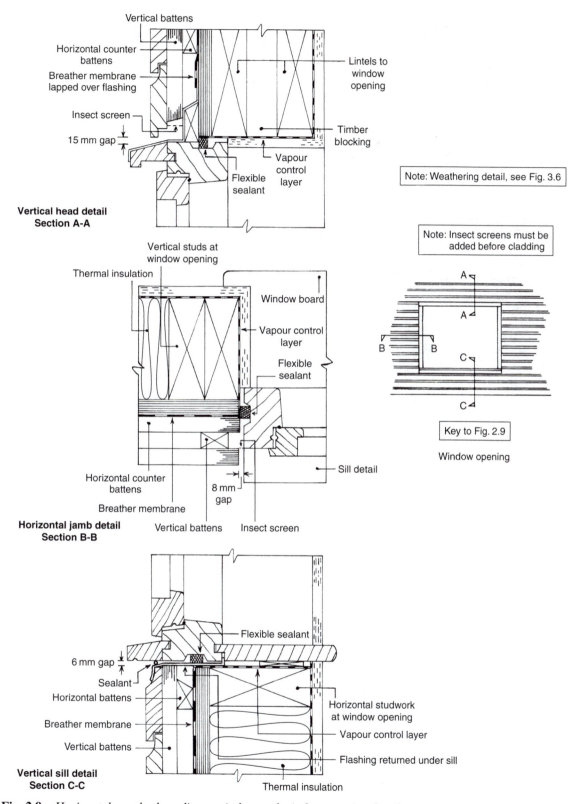

Vertical head detail Section A-A

Vertical battens

Horizontal counter battens

Breather membrane lapped over flashing

Insect screen

15 mm gap

Lintels to window opening

Timber blocking

Vapour control layer

Flexible sealant

Note: Weathering detail, see Fig. 3.6

Note: Insect screens must be added before cladding

Horizontal jamb detail Section B-B

Vertical studs at window opening

Thermal insulation

Window board

Vapour control layer

Flexible sealant

Horizontal counter battens

Breather membrane

Vertical battens

Insect screen

8 mm gap

Sill detail

Key to Fig. 2.9

Window opening

Vertical sill detail Section C-C

Flexible sealant

6 mm gap

Sealant

Horizontal battens

Breather membrane

Vertical battens

Horizontal studwork at window opening

Vapour control layer

Flashing returned under sill

Thermal insulation

Fig. 2.9 *Horizontal weatherboarding – window and window opening details*

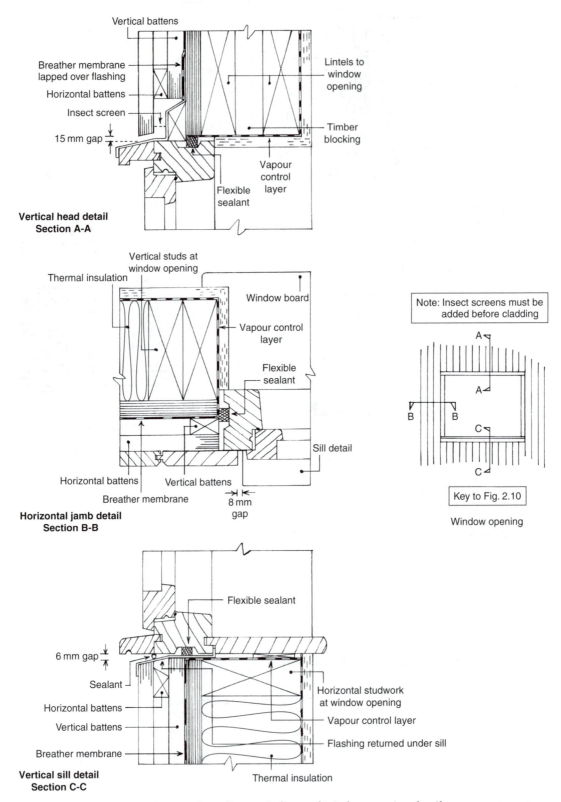

Fig. 2.10 *Vertical tongue-and-groove boarding – window and window opening details*

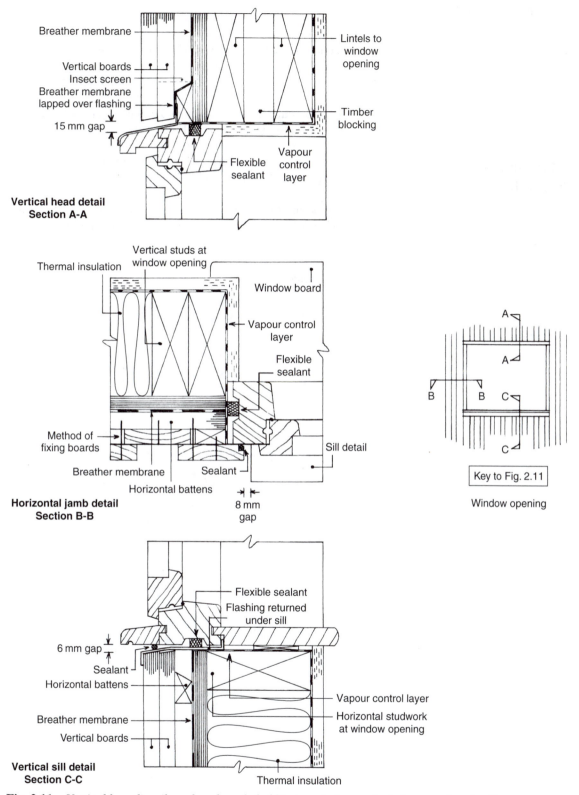

Fig. 2.11 *Vertical boarding (board-on-board cladding) – window and window opening details*

Two stainless steel nails are used with each shingle to form a single layer to include a double layer at the base for extra protection – ferrous metal nails should not be used (see Volume 1, section 1.10.3g).

Joints must be staggered with a gap of 5 mm between each shingle to allow for moisture movement (see Fig. 2.1e).

2.4.2 Shakes (Fig. 2.12f)

In the case of shakes, again the log is cut to the required lengths (similar to shingles), and quarter sawn (Fig. 2.12a).

However, before they can be tapered, the blocks are 'split' using a cleaver (Fig. 2.12d) to produce a 'rough' cut finish. They are then taper

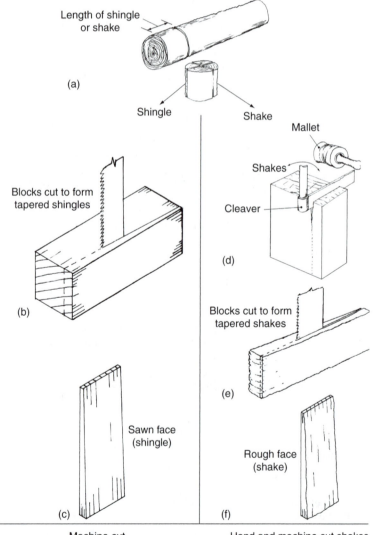

Fig. 2.12 *Producing shingles and shakes*

sawn (Fig. 2.12e) to produce a shake with the 'rough' face on the outside (Fig. 2.12f).

Shakes are usually 100 to 355 mm in width and 610 mm long tapering down to 4 mm.

2.4.2.1 Fixing

Similar to shingles, shakes are fixed by securing stainless steel nails to horizontal grounds.

2.5 Plywood cladding

Plywood must have been manufactured for exterior use (see BS EN 636: 2003, Plywood specifications; WBP, see Vol. 1, section 4.1.2) and should be not less than 9.5 mm thick. It may be used in strip form (as weatherboard) or sheet (panel) form (Fig. 2.13b).

2.5.1 Detailing

A typical corner detail of ply cladding is shown in Figs 2.13a and 2.13b together with a suitable corner cover lath.

Figure 2.13b shows how a vertical ply panel abutment can be formed – edge exposure to the weather should always be reduced to a minimum.

Figure 2.14 shows detailing around a window opening. As with other timber cladding, back-venting is a requirement – it can be achieved in a similar way to solid timber cladding (see Fig. 2.6).

2.5.2 Fixing plywood cladding

Nails used for fixing the cladding should be hot-dip galvanized or corrosion-resistant, for example stainless steel, silicon bronze, or copper. Aluminium nails are unsuitable when the timber has been treated with certain wood preservatives.

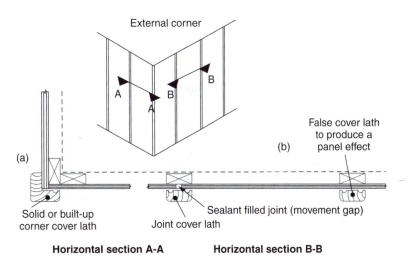

Fig. 2.13 *Vertical and corner joints for plywood panels*

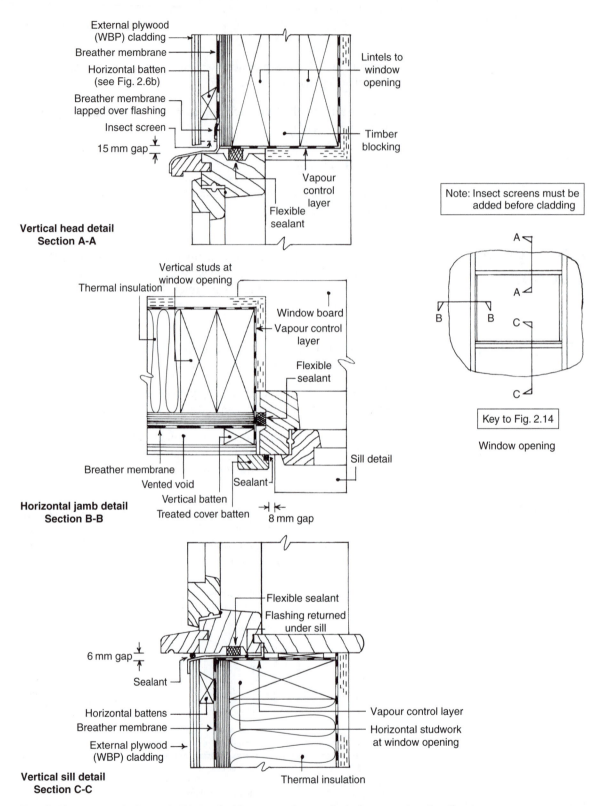

External plywood (WBP) cladding

Breather membrane

Horizontal batten (see Fig. 2.6b)

Breather membrane lapped over flashing

Insect screen

15 mm gap

Lintels to window opening

Timber blocking

Vapour control layer

Flexible sealant

Vertical head detail
Section A-A

Thermal insulation

Vertical studs at window opening

Window board

Vapour control layer

Flexible sealant

Breather membrane

Vented void

Vertical batten

Treated cover batten

Sealant

Sill detail

8 mm gap

Horizontal jamb detail
Section B-B

Note: Insect screens must be added before cladding

Key to Fig. 2.14

Window opening

Flexible sealant

Flashing returned under sill

6 mm gap

Sealant

Horizontal battens

Breather membrane

External plywood (WBP) cladding

Vapour control layer

Horizontal studwork at window opening

Thermal insulation

Vertical sill detail
Section C-C

Fig. 2.14 *External plywood (WBP) cladding – window and window opening details*

Where extra-good holding power is important, improved nails (Volume 1, Table 12.1) can be used. As a general guide, nails should have a length of two and a half times the thickness of the board they are to fix (annular-ring nails need only be twice this thickness).

Nails should be positioned as shown in Fig. 2.3, to help counter the effect of moisture movement without the boards splitting. Face fixing with round rustproof wire nails is common with weatherboarding, board-and-batten, and board-on-board cladding. Matchboarding can be secret nailed (the nail head being concealed by the groove in the board) or fixed with proprietary metal clips, which are concealed by the board that follows.

2.6 Surface finishes to timber cladding

The use of timber preservatives requires proper safety precautions when applying, as most are classified as inflammable and harmful. The types in general use include organic-solvent and water-borne preservatives (see Volume 1, section 3.3). As with a lot of materials they may contain harmful chemicals, so it is advisable to consult the COSHH regulations before use. Furthermore, there must be adequate ventilation, the provision and use of personal protective equipment, and strict following of the manufacturer's instructions. (See BS 1282: 1999, Wood preservatives. Guidance on choice, use and application.)

Surface finishes, i.e. paints, varnishes and woodstains can be classified as low to high build surface coatings. For example:

(i) Paint – High-build, will obscure the surface below making it opaque.
(ii) Varnishes – High-build, allows the underlying surface to be seen (transparent) and provide protection.
(iii) Woodstains – Low to High build, provide a coloured surface finish which can be semi-transparent or opaque.
Low build – offers minimal surface protection.
Medium build – offers adequate protection.

Finishes can have decorative and weather-protective qualities, yet be permeable to water vapour (micro-porous); that is to say, if water vapour enters the wood from behind the cladding (most should be vented away) or moisture enters via surface jointing, etc., it can escape through the permeable (breather) surface. If an impermeable paint or varnish film is used as outer surface protection and moisture enters the wood via small cracks, etc., it could become trapped and cause a breakdown of the surface film. Typical examples of film failures are bubbling, flaking, and splitting paintwork. With clear varnishes, the timber below could become discoloured with blue-stain fungi, etc.

Suitable micro-porous finishes can be grouped as:

(a) exterior emulsion paints;
(b) exterior wood stains;
(c) certain alkyd paints.

2.6.1 Exterior emulsion paints

These are water-based acrylic paints producing an opaque coloured 'silk' finish. Application methods vary – some manufacturers may specify two brush coats, others three. Surfaces may last as long as five years without retreatment; however, this will depend greatly on the amount and type of exposure, for example atmospheric pollution.

2.6.2 Exterior wood stains

These have a water-repelling non-opaque (translucent) matt to semi-gloss finish and are available in many light-fast colours. These stains generally protect wood against blue-stain and mould growth. Application usually involves a minimum of two brushed coats.

> Note: *It would appear that both the above groups perform best when they are applied to sawn timber (unplaned).*

2.6.3 Alkyd paints

Although not generally associated with cladding, alkyd paints should also be mentioned. These paints can be classified as:

(a) general-purpose alkyds – non-breather-type impermeable paints, available in a high-gloss finish.
(b) exterior-quality alkyds – low-gloss paints with low permeability (micro-porous), suitable for outside use but usually confined to coating the outer surfaces of exterior joinery, i.e. doors, windows, etc.

2.6.4 Plywood finishes

Plywood may be available as a pre-finished panel which has received a heavy coating of synthetic resin – possibly covered with a layer of white or coloured mineral chippings (see Volume 1, section 4.1.2).

2.7 Other claddings

Tempered hardboard, exterior-grade cement-bonded particle board, and non-combustible boards (see Volume 1, Chapter 4) can all be fixed in a similar way to plywood, i.e. on preservative-treated back battens. Surface treatment will be similar to that for timber.

Plastics are usually used in the form of weatherboard, and available in a variety of finishes including simulated timber, shown in Fig. 2.15, together with fixing details. Corner and edge detailing usually involves

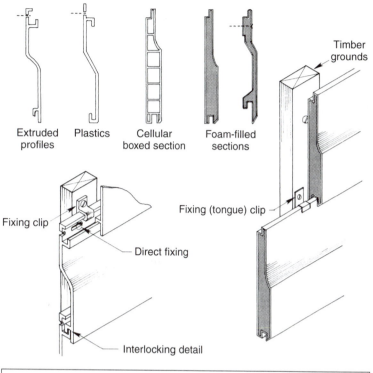

Fig. 2.15 *Plastics (uPVC) weatherboarding*

Note: Sections for edging and jointing internal and external angles are available

using special extruded sections supplied by the manufacturer of the main profile.

Masonry – brickwork, stonework, and blockwork – is very popular as a cladding to timber-framed houses. Masonry is attached to the load-bearing timber structure by means of flexible metal ties that allow for any differential movement between the dissimilar materials – examples are shown in Chapter 1, Figs 1.7 and 1.9.

The use of cement rendering (see Table 2.1 and Fig. 2.1f) (a mixture of sand, cement, and sometimes lime (or substitute), with water to make a mortar) as a cladding relies on sheets of expanded metal stapled across a framework of timber to key, hold, and reinforce the rendering when set. These days, proprietary metal lath backings are available in the form of galvanized welded wire mesh, reinforced with stainless steel wire. A sheet of perforated absorbent stiff paper is incorporated within the fabric, and a breather paper may also be included. The lath is stapled with stainless steel staples every 150 mm to vertical preservative-treated battens fixed at 400 mm to 600 mm centres. A ventilated space of at least 25 mm deep should be left behind the rendering.

Tile hangings – tiles or shingles – are lapped as roof tiles and double-nailed to preservative-treated battens. Because the battens will be fixed horizontally, notches may be cut or gaps left along their run – or counter battens used as previously mentioned with vertical matchboard – to allow for full back-venting of the cavity.

Metal cladding is usually available as a proprietary cladding system, in which case manufacturers will supply all the necessary fixings and fixing details.

References

BS 1282: 1999, Wood Preservatives. Guidance on choice use and application.

BS 4072: 1999, Copper/Chromium/Arsenic – Preparations for wood preservatives.

BS 4261: 1999, Wood Preservation, Vocabulary.

BS 5589: 1989, Code of practice for wood preservation.

BS 5268–5: 2005, Structural use of timber, code of practice for the preservative treatment of structural timber.

BS 5707: 1997, Specification for preparations of wood preservatives in organic solvents.

BS 5534: 2003, Code of practice for slating and tiling (including shingles).

BS 476–3: 2004, Fire tests on building materials and structures. Classification and method of test for external fire exposure to roofs.

BS 7956: 2000, Specification for primers for woodwork.

BS 6150: 2006, Code of practice for painting of buildings.

BS 476–7: 1997, Fire tests on building materials. Surface spread of flames.

BS 8414–1: 2002, Fire performance of external cladding systems.

BS EN 634–2: 1997, Cement-bonded particle board, specification.

BS EN 633: 1994, Cement-bonded particle boards, definition.

BS EN 636: 2003, Plywood specifications, Fire safety: 2000.

Building Regulations Approved Document B1, Fire Spread: 2000.

3 Windows

The main purpose of a window is to allow natural light to enter a building yet still exclude wind, rain, and snow. Windows are also a means of providing ventilation when the glazed area is made to open and trickle vent (Fig. 3.64) within their frame. An openable 'light' (a single-glazed unit of a window) is called a 'casement' or 'sash' – hence the term 'casement window' or 'sash window'.

3.1 Window profiles

Timber components of the window need to be profiled externally in such a way as to prevent water penetrating the building through small gaps by the process of capillarity (see section 3.1.2). Failure to do so could result in a high moisture content of timber and its eventual breakdown by moisture-seeking fungi, not to mention the many other structural and environmental effects associated with dampness (as featured in Volume 1, section 2).

Modern methods of window construction have mostly overcome this problem by the introduction of various means of weathering to the frame (Fig. 3.10).

Most building materials are porous (i.e. contain voids or pores), thus allowing moisture to travel into or through them. Moisture may enter these materials from any direction, even the underside, and travel upwards as if defying the laws of gravity.

Upward movement of moisture is responsible for rising damp and is caused by a force known as 'capillary' force, which in the presence of surface tension produces an action called 'capillarity'.

3.1.1 Surface tension (Fig. 3.1)

The microscopic molecules of which water is composed are of a cohesive (uniting or sticking together) nature which, as a result of

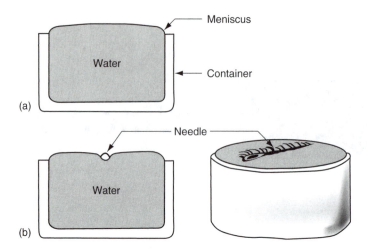

Fig. 3.1 *The effect of surface tension*

their pulling together, produces the apparent effect of a thin flexible film on the surface of water. The effect is known as 'surface tension'. As can be seen in Fig. 3.1, the reluctance of this 'skin' to be broken can be shown by filling a container with water just below its brim, then slowly adding more water until a meniscus (curved surface of liquid) has formed (Fig. 3.1a). Its elasticity can be demonstrated further by carefully floating a small sewing needle on its surface. Figure 3.1b shows how a depression is made in the 'skin' by the weight of a needle. If, however, surface tension is 'broken' by piercing the 'skin' the needle will sink.

3.1.2 Capillarity (Fig. 3.2)

A clear clean glass container partly filled with water will reveal a concave meniscus where the water has been drawn up the sides of the glass (Fig. 3.2a). This indicates that a state of adhesion (molecules attachment of dissimilar materials) exists between the glass and water molecules and at this point is stronger than the cohesive forces within the water.

Figure 3.2b shows what happens when a series of clean glass tubes is stood in a container of water. The water rises highest in the tube with the smallest bore (hole size), indicating that the height reached by capillarity is related to the surface area of water in the tube. It can therefore be said that, if the surface area of water is reduced or restricted, the downward pull due to gravity will have less effect, whereas upward movement will be encouraged by surface tension.

A similar experiment can be carried out by using two pieces of glass as shown in Fig. 3.2c. This method produces a distinct meniscus which can be varied by making the gap at the open end wider or narrower. It should now be apparent that, for capillary force to function,

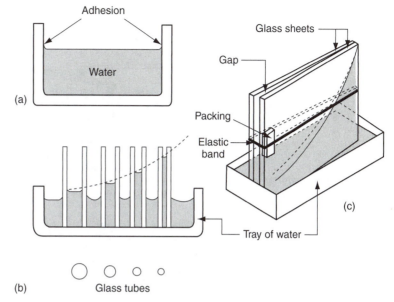

Fig. 3.2 *The effect of capillarity*

all that is needed is water with its cohesive properties and a porus material, or a situation providing close 'wettable' surfaces which encourage adhesion.

3.1.3 Preventative measures against the entry of moisture

Knowledge of how water acts in certain situations, and reacts towards different materials, enables a building and its components to be designed with built-in water and moisture checks. For example, the common method of preventing moisture from rising from the ground into the structure is by using a horizontal damp-proof course (DPC) like those shown in Volume 2, Fig. 5.1. Penetrating damp has also been avoided by leaving a cavity between outer and inner wall skins. However, a vertical DPC would be used where both skins meet around door and window openings.

There are many other places – particularly the narrow gaps left around doors and sashes – which provide ideal conditions for capillary action if preventative steps are not taken. The first line of defence should be to redirect as much surface water away from these areas as possible, by recessing them back from the face wall and encouraging water to drip clear of the structure.

Figure 3.3 shows how a 'throat' (groove) cut into the underside of over-hangs, i.e. thresholds, window sills, drip moulds, etc., interrupts the flow of water by forcing it to collect in such a way that its increase in weight results in the drip being formed. Further examples of throat-ings are shown in Figs 3.6, 3.10, 3.15 and 3.16.

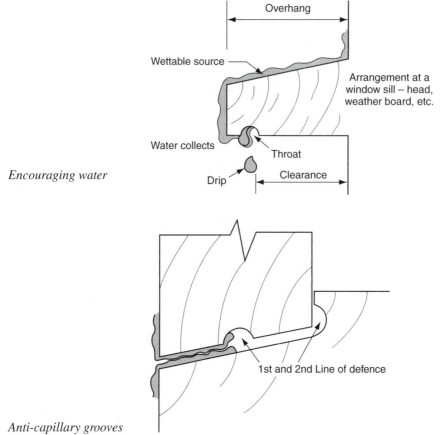

Fig. 3.3   *Encouraging water to drip*

Fig. 3.4   *Anti-capillary grooves*

It is inevitable that water will find its way around those narrow gaps, so anti-capillary grooves, or similar, are used. Figure 3.4 shows how these grooves work, and Fig. 3.6 illustrates how they are included in window design.

3.2 Types of windows

Most standard windows are made from softwood, usually European Redwood (*pinus sylvesttis*) or European whitewood (*picea abies* or *abies alba*), both non-durable, which, at the manufacturing stage, is treated with wood preservative before being coated with a wood primer (base coat). The method of preservative treatment will depend on the stage at which treatment is carried out; for example, after components have been machined or at the assembled staged (see Volume 1 Chapter 3; also CP 153: part 2: 1970, Table 3).

Some hardwood timbers may also be used for extra durability, including Idigbo (*terminalia ivorensis*), European Oak (*quercus robur*), Meranti/Red Lauan (shorea spp), Sapele (*entandrophragma cylindricum*) (see Volume 1, Table 1.15), and Brazilian Cedar (cedrela spp).

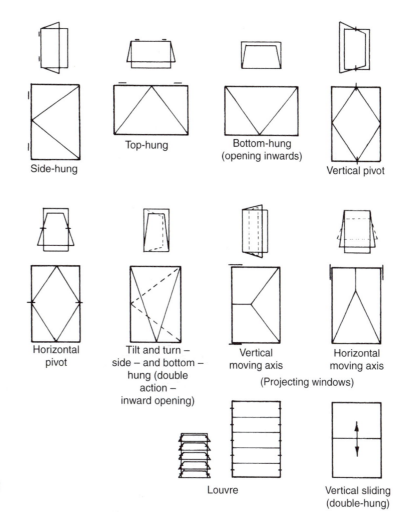

Side-hung

Top-hung

Bottom-hung
(opening inwards)

Vertical pivot

Horizontal
pivot

Tilt and turn –
side – and bottom –
hung (double
action –
inward opening)

Vertical
moving axis

Horizontal
moving axis

(Projecting windows)

Louvre

Vertical sliding
(double-hung)

Fig. 3.5 *Diagrammatic method of showing opening lights*

An opening light may operate in many different ways, for example, it may be swung, tilted, turned, or slid; and in some cases it may combine more than one of these operations to facilitate cleaning or draught prevention. Figure 3.5 has been included to simplify identification.

3.2.1 Single-light casement windows

Generally, the frame will consist of a head, jambs, and sill (or cill) for a simple single-light casement window, but other types, such as multi-light casement windows, include Transoms and Mullions (see section 3.2.4).

The casement is constructed with a top rail, a pair of stiles and a bottom rail – glazing bars may also be included to reduce the glazed area as well as adding an aesthetic design to the window.

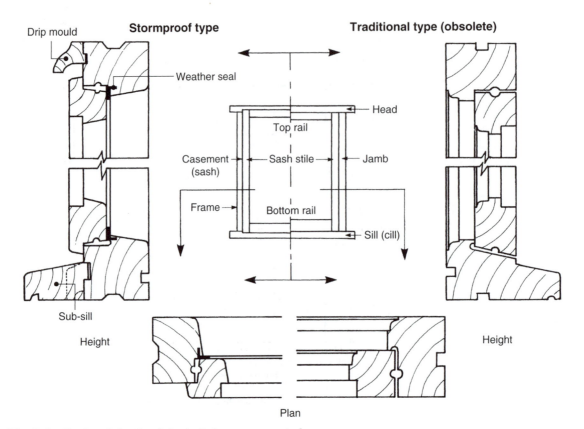

Fig. 3.6   *Sectional details of single-light casement windows*

Figure 3.6 shows details of two types of casement windows:

- the obsolete 'traditional' type, which housed its casement fully within its frame; and
- the more modern 'stormproof' type which, because of its double-rebate system, projects slightly outside the window frame – this is brought about because both the frame and the casement are rebated – thereby giving much better weather protection as a whole, and the added benefit of increased natural light.

Because the rebate in the frame of the 'stormproof' type is only half the casement width, direct glazing or deadlight (non-openable) can be incorporated into the window frame without the need of a casement.

3.2.2 Construction

Construction methods will depend mainly on the type of window and the sectional profile of its members. Traditional sections (and stormproof sections to BS 644: 2003, Timber windows) are shown in detail in Fig. 3.6 (Fig. 3.10 shows how these sections can be modified). The important feature of all these sections is how the grooves are formed to provide protection against the entry of moisture by capillary action (see Fig. 3.4).

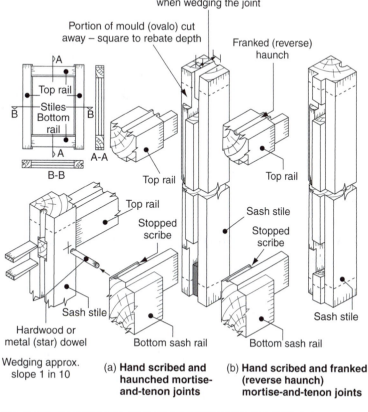

Width reduced for haunch – possible
area of weakness (top and bottom joints)
when wedging the joint

Portion of mould (ovalo) cut
away – square to rebate depth

Franked (reverse)
haunch

Top rail

Stiles
Bottom
rail

A-A

B-B

Top rail

Top rail

Top rail

Sash stile

Top rail

Stopped
scribe

Stopped
scribe

Stopped
scribe

Sash stile

Hardwood or
metal (star) dowel

Bottom sash rail

Bottom sash rail

Sash stile

Wedging approx.
slope 1 in 10

(a) **Hand scribed and
haunched mortise-
and-tenon joints**

(b) **Hand scribed and franked
(reverse haunch)
mortise-and-tenon joints**

Fig. 3.7 *Joints using traditional
sash stile and sash rail*

Note:
1. The term 'sash' stile or rail is derived from the wood sectional profile
    used to make up sashes for sash windows (see section 3.2.8). The same
    wood sections, however, are used to construct traditional casements.
2. Mortise-and-tenon joints: the use of a 'franked' haunch (as shown in
    detail 'b') on shallow rebate/moulded small sectioned stock reduces the
    risk of splitting the stile during wedging and can produce a stronger joint

Window frames should be held together at the corners with either
combed joints (multiple corner bridles) with not less than two tongues,
or mortise-and-tenon joints. The tongues or tenons should not be less
than 12 mm in thickness. Where horns are required for building-in purposes, they should not be less than 40 mm in length. Joints should be
glued, with a suitable synthetic adhesive, and be held together with
wedges and/or pegged with hardwood or metal star dowels (Fig. 3.7).

Casements should be joined by using combed or mortise-and-tenon
joints, glued (as above), and pegged and/or wedged (depending on
the type – bridle joints cannot be wedged). Traditional casement
joints are shown in Fig. 3.7. Stormproof casements details are shown
in Fig. 3.12.

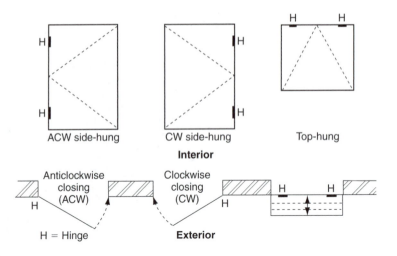

Fig. 3.8 *Opening casements*

3.2.3 Hanging opening casements

Sashes can be hung in one of the three ways shown in Fig. 3.8, where the V indicates the hinging side and that the knuckle (Fig. 3.58) of the hinge is to that face. (Three different types of hinge are shown in Fig. 3.58, together with their respective fixing details.)

The screws used to fix hinges will be exposed to varying amounts of strain, particularly those used on the top hinges on side-hung sashes, and those on sashes which are top hung, which will constantly be subjected to *withdrawal*.

Because openable casements must provide varying amounts of ventilation a multi-positioned casement stay similar to the one shown in Fig. 3.58 is used. This also gives a means of securing the sash when closed. Side-hung casements also use a window fastener (Fig. 3.58) fixed midway to the closing side stile and jamb.

Glass is not usually fixed until the window is built into the structure, but its weight must be taken into account with regard to hinge and screw sizes. The glass also plays an important part in holding the casement square (with regard to setting blocks see the section on glazing – section 3.8).

3.2.4 Multi-light standard (stormproof) casement windows

Introducing more than one light into a window frame will mean its division by other members, namely mullions (intermediate vertical members) and/or transoms (intermediate horizontal members). The possible combinations are many and vary according to the number and size of lights and whether they are openable or dead. Figure 3.9 shows how they may appear, but to fully appreciate all the possibilities, including combination with door frames, manufacturers' catalogues should be consulted.

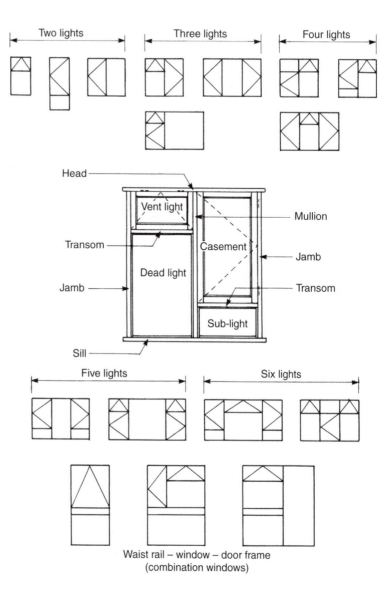

Fig. 3.9 *Variations in casement window design*

Waist rail – window – door frame
(combination windows)

Figure 3.10 shows joint locations and section details of a standard three-light casement window. Provision has been made for a resilient weather-strip within the casement rebate of the frame. Window frame joints for this type of window (Fig. 3.11) are usually made with a mortise and tenon – glued, wedged, and/or pinned through the tenon from the inside face of the frame with metal or wood dowel. For reasons of economy, some manufacturers use scribed butt joints held together with glue and screws in place of the mortise and tenon. Casement joints between rails and stiles are combed joints, glued and pinned with metal or wood dowel from the inside face of the case-ment. Generally the faces of the casement would reveal the stiles run-ning the full height of the casement opening, with the rails butting up

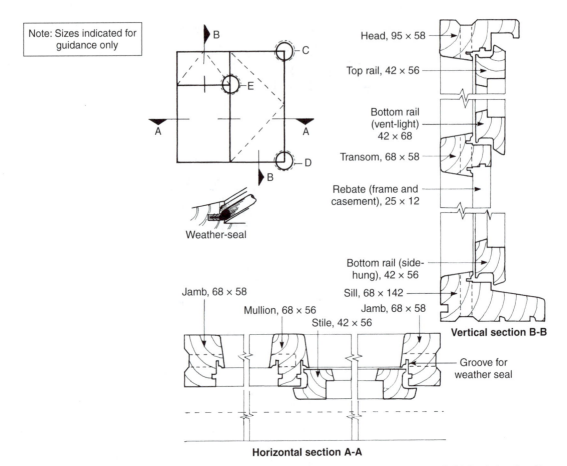

Note: Sizes indicated for guidance only

Head, 95 × 58

Top rail, 42 × 56

Bottom rail (vent-light) 42 × 68

Transom, 68 × 58

Rebate (frame and casement), 25 × 12

Bottom rail (side-hung), 42 × 56

Sill, 68 × 142

Jamb, 68 × 58

Vertical section B-B

Weather-seal

Jamb, 68 × 58

Mullion, 68 × 56

Stile, 42 × 56

Groove for weather seal

Horizontal section A-A

Fig. 3.10 *Details of a three-light improved 'standard' stormproof window (see Fig. 3.11 for joint details at C, D and E)*

to them as shown in Fig. 3.12. However, end grain is exposed to the elements and regular maintenance is required at these points.

3.2.5 Fixing a casement window to the structure

Window frames can either be built into the fabric of the structure as the building progresses – in a similar manner to door frames as shown in Fig. 4.48 – or more commonly now fixed with screws and plugs into a pre-formed opening after the main structure is built.

Approved Document L1 and L2, Conservation of Fuel and Power: 2002 requires that all types of construction must have improved thermal insulation. This applies especially to the installation of new windows and external door frames if the problems of mould growth and condensation are to be overcome. Problems arise through thermal bridging where the cavities have been closed using traditional methods and around openings situated in external walls. A thermal cavity closer is now required in these locations – which also act as a DPC.

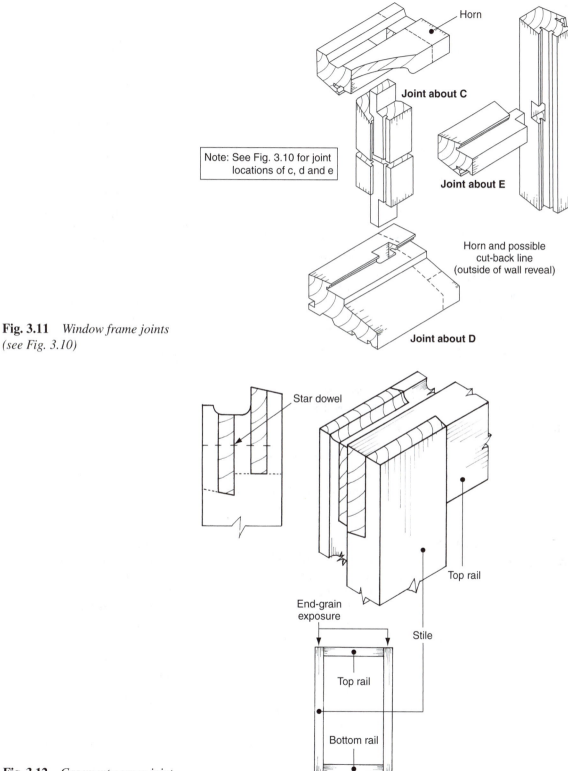

Fig. 3.11 *Window frame joints (see Fig. 3.10)*

Fig. 3.12 *Casement corner-joint (combed) arrangements*

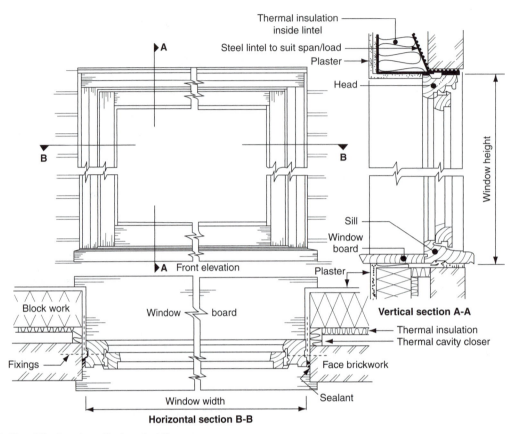

Fig. 3.13   *Window installation details*

Thermal cavity closers are available from proprietary manufacturers in a variety of shapes and sizes to fit any situation. Figure 3.13 shows a vertical section of how a typical window is incorporated into the main outer fabric of the house.

3.2.6 Casement bay windows

A bay is an integral part of a building that projects outwards beyond the main building line. It may be square, rectangular, splayed, or segmental in plan, and it may extend one or more storeys above the ground. If only the window itself projects, without any substructure to ground level, it is called an 'oriel' window. Bay windows are built from a number of standard or high-performance windows joined together at their corners to produce the following named types:

(a) *Square bays* (sometimes called 90° bays) (Fig. 3.14a): the fronts of these bays, unlike the ends, are often made up of more than one frame. Bay corners (angular mullions) may be constructed by one of three methods:
   (i)   a solid corner post (traditional);
   (ii) a mitred section with either an inset fillet (tongue) or a face cover;
   (iii) a solid corner infill.

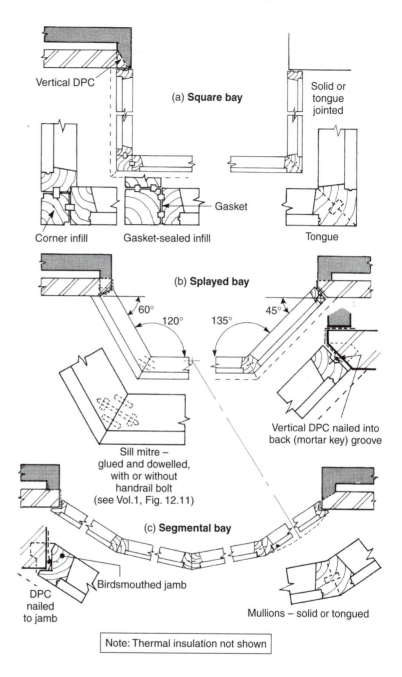

Vertical DPC

(a) **Square bay**

Solid or tongue jointed

Gasket

Corner infill   Gasket-sealed infill   Tongue

(b) **Splayed bay**

60°   45°
120°   135°

Vertical DPC nailed into back (mortar key) groove

Sill mitre – glued and dowelled, with or without handrail bolt (see Vol.1, Fig. 12.11)

(c) **Segmental bay**

Birdsmouthed jamb

DPC nailed to jamb

Mullions – solid or tongued

Fig. 3.14 *Bay window shapes and corner construction*

Note: Thermal insulation not shown

(b) *Splayed bays* (Fig. 3.14b): these have a similar construction to square bays, except that the sides are splayed – usually to an angle of 45° or 60°.

(c) *Segmental bays* (Fig. 3.14c): each bay segment may be made up of three to seven window frames to make further segmental subdivisions.

These frames usually have a curved sill, which is why this type of bay is sometimes known as either a 'circular' or a 'shallow circular' bay. Figure 3.14 also shows how these bays meet up with wall reveals.

3.2.7 High-performance casement windows

Note: *The design principles which follow also apply to top-hung casements, but vent lights generally occur only within the framework of energy-saving types.*

For the purpose of this chapter, an energy-saving type of window is one which obtains its efficiency from an 'all-round' draught- and weather-seal and has the capability of being glazed with a sealed double-glazed unit of a total thickness not exceeding 24 mm, as shown in Fig. 3.15, unless a stepped-edge type is used (Fig. 3.49).

Figure 3.16 shows member details of a high-performance window. Most of these windows are in hardwood. Some models may have built-in ventilation in the form of a duct mortised into the top rail of the case-ment, which can be regulated from inside the building (Fig. 3.64). Casement rebates often accommodate sealed double-glazed units of up to 24 mm thickness. Special sliding hinges (friction-stay types) permit easy cleaning from inside the building – details of these hinges and special latching and locking mechanisms are shown in Figs 3.59, 3.62, and 3.63.

The term 'high-performance window' can be misleading, but generally means a window which has one or more energy-saving feature within its design, such as weather-seals, thermal insulation, and 'Low E' glazing (see section 3.5). Many have integral ventilation and security fasteners. Most have a type of hinge mechanism which enables openable lights to be cleaned from inside the building.

High-performance windows are generally made from a hardwood or from the softwood Douglas fir (*Pseudotsuga menziesii*) – both should be moderately durable. The former may be better and, provided sapwood is excluded, preservative treatment may be optional. However, selected European redwood (*Pinussylvestris*), which is non-durable, may be used, in which case preservative treatment would be necessary.

3.2.8 Double-hung sash windows

Firstly, let us differentiate between a sash window and a casement window. Both are designed to house the glazing, but opening case-ments are hinged, whereas sashes slide open.

Double-hung sash windows consist of two vertical sliding sashes, one offset above the other. Both sashes are counterbalanced at each side either by a weight, suspended by a line (cord or chain), or by a spring balance mechanism (see section 3.2.14). Figure 3.17 shows the basic principle of using a weight to counterbalance the weight of the sash and its glass.

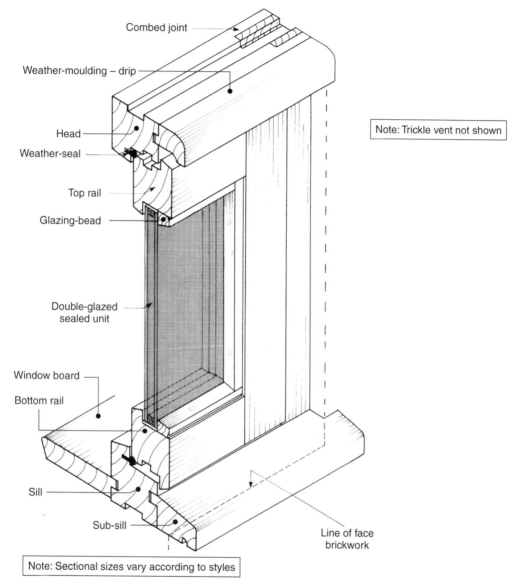

Combed joint

Weather-moulding – drip

Head

Weather-seal

Top rail

Glazing-bead

Double-glazed
sealed unit

Window board

Bottom rail

Sill

Sub-sill

Note: Trickle vent not shown

Line of face
brickwork

Note: Sectional sizes vary according to styles

Fig. 3.15 *Energy-saving casement window*

There are two main types of window: the traditional cased windows
with 'cased' or 'boxed' frames (the case houses the weights), and the
more modern types which have 'solid' frames with a spiral spring balance either attached or housed in them.

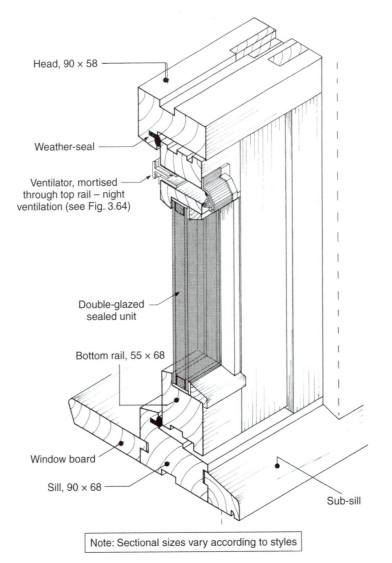

Head, 90 × 58

Weather-seal

Ventilator, mortised
through top rail – night
ventilation (see Fig. 3.64)

Double-glazed
sealed unit

Bottom rail, 55 × 68

Window board

Sill, 90 × 68

Sub-sill

Note: Sectional sizes vary according to styles

Fig. 3.16 *High-performance casement window*

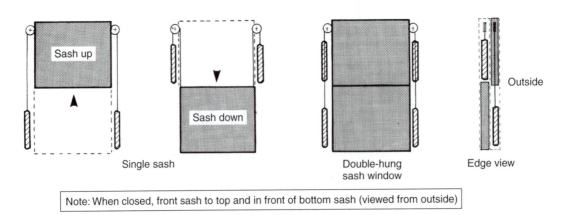

Sash up

Sash down

Single sash

Double-hung
sash window

Outside

Edge view

Note: When closed, front sash to top and in front of bottom sash (viewed from outside)

Fig. 3.17 *Counterbalancing sashes with weights*

3.2.9 Cased frames

Figures 3.18 and 3.19 show the arrangement of components and how they are assembled and joined together. Pulley stiles are trenched into the head and sill and back-wedged within the sill housing. The inner and outer linings are grooved, to receive the tongues cut into the edges of the pulley stiles and head. Right-angle joints between the head and linings are stiffened with glue blocks.

Detachable 'pockets' are cut into both pulley stiles (Fig. 3.20) to allow access to the weights and sash lines. Two pulley wheels are mortised

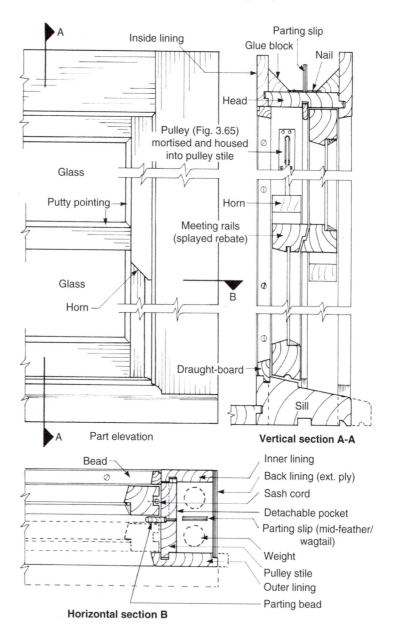

A

Inside lining

Parting slip

Glue block

Nail

Head

Pulley (Fig. 3.65) mortised and housed into pulley stile

Glass

Putty pointing

Horn

Meeting rails (splayed rebate)

Glass

Horn

B

Draught-board

Sill

A   Part elevation

Vertical section A-A

Bead

Inner lining

Back lining (ext. ply)

Sash cord

Detachable pocket

Parting slip (mid-feather/ wagtail)

Weight

Pulley stile

Outer lining

Parting bead

Horizontal section B

Fig. 3.18 *Part double-hung sash window with cased frame*

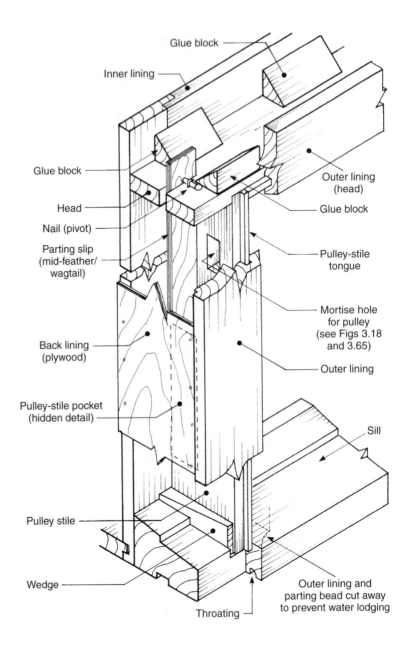

Fig. 3.19 *Sectional constructional cut-away details of a cased frame*

into each stile, with their centres not more than 115 mm or less than 90 mm from the head.

The sashes are held in check by the linings on the outer side and by the bead on the inner. At sill level, an upstand (draught-bead) may be incorporated in the design, which should (theoretically) allow a gap to be left between meeting rails for ventilation purposes without creating a draught at sill height. Sashes are held apart with parting bead, which is housed into both pulley stiles and the head.

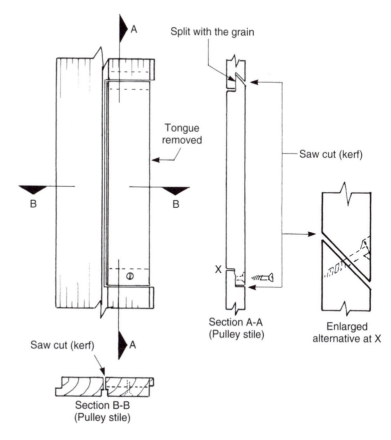

Fig. 3.20 *Forming a detachable pocket*

Although some are left open, the back of the cased frame should be lined with exterior-quality plywood to prevent mortar droppings, dust, etc. from entering the box and restricting the distance the weight can travel. (If the top-sash weight were not allowed to drop its full distance, the sash would not close.)

A parting slip (or 'wagtail' or 'mid-feather') should be mortised through the head of each side and held up with a peg or nail, which allows it to swing freely (its purpose is to separate the weights). On tall windows, weights tend to swing when the sashes are being opened – the parting slip prevents the weights from touching one another.

3.2.10 Solid frames

Figure 3.21 shows vertical and horizontal sections through the frame and sashes. Alternative methods of housing the spiral spring balances are also shown in Fig. 3.24.

Joints between jambs, head, and sill may be combed-jointed or mortised-and-tenoned. Outer linings are nailed and glued to the jambs

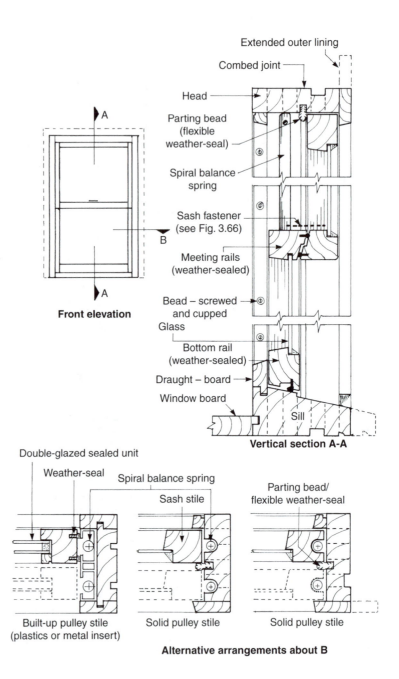

Front elevation

Extended outer lining

Combed joint

Head

Parting bead (flexible weather-seal)

Spiral balance spring

Sash fastener (see Fig. 3.66)

Meeting rails (weather-sealed)

Bead – screwed and cupped

Glass

Bottom rail (weather-sealed)

Draught – board

Window board

Sill

Vertical section A-A

Double-glazed sealed unit

Weather-seal   Spiral balance spring

Sash stile

Parting bead/ flexible weather-seal

Built-up pulley stile (plastics or metal insert)

Solid pulley stile

Solid pulley stile

Alternative arrangements about B

Fig. 3.21 *Double-hung sash window with solid frame*

and head. Parting beads may incorporate a form of weather-seal, and staff beads are sometimes omitted to allow the sashes to tilt inwards for easy cleaning and to be removed from their frames for maintenance purposes, i.e. painting, etc. Some proprietary solid-framed sash windows are categorized as having high performance – because of the facilities already mentioned and their ability to accept sealed double-glazed units.

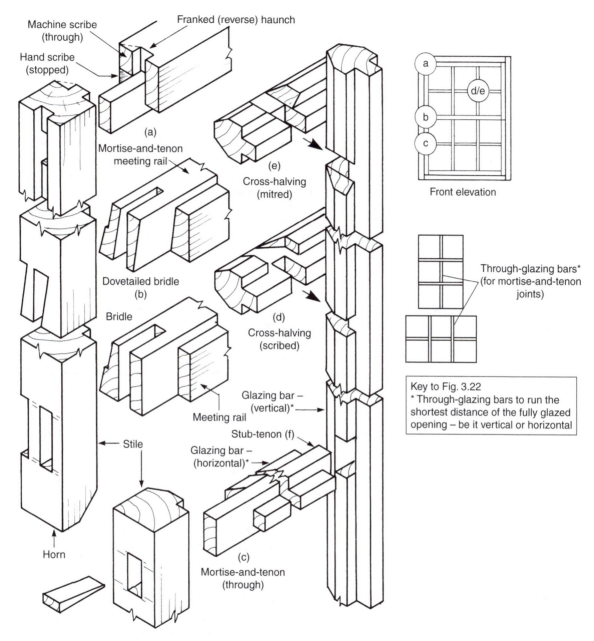

Machine scribe (through)

Hand scribe (stopped)

Franked (reverse) haunch

(a)

Mortise-and-tenon meeting rail

Dovetailed bridle (b)

Bridle

Stile

Horn

Meeting rail

(c)

Mortise-and-tenon (through)

(e)

Cross-halving (mitred)

(d)

Cross-halving (scribed)

Glazing bar – (vertical)*

Stub-tenon (f)

Glazing bar – (horizontal)*

Front elevation

Through-glazing bars* (for mortise-and-tenon joints)

Key to Fig. 3.22
* Through-glazing bars to run the shortest distance of the fully glazed opening – be it vertical or horizontal

Fig. 3.22 *Traditional sash joints*

3.2.11 Sashes

Corners are scribed, with either a combed joint (Fig. 3.12) or a mortise-and tenon joint (Fig. 3.22a). Through tenons may be used where horns are to be retained, otherwise a haunch should be used. Where the backs of stiles are grooved to receive cords or a spring balance, a franked haunch (see Fig. 3.22a) – also known as a 'reverse haunch' – should be used, otherwise the small amount of wood left between the haunch and the groove would easily split away. All joints should be either glued and pinned or glued and wedged and pinned, depending on their type.

Meeting rails are splayed (Figs 3.18 and 3.22) and rebated to prevent draught and as a security measure. Alternative joints are shown in Fig. 3.22b.

If glazing bars (sash bars) are used to divide the window into smaller panes, they should be scribed and tenoned into mortise holes within the sash (Fig. 3.22c). Vertical members should be through-tenoned; horizontal members stub-tenoned. Where glazing bars cross, a scribed cross-halving joint (Fig. 3.22d) or a mitred cross-halving joint (Fig. 3.22e) – usually with hardwood – may be used, or horizontal cross-members may be scribed and stub-tenoned into vertical members (Fig. 3.22f).

3.2.12 Sash counter-balances

To ensure that the sashes slide easily up or down, with minimum effort, weights or proprietary spring-balances are used as compensators. Lead was traditionally used for the weights, but nowadays they are generally made of cast-iron, and can be either round or square in section. To determine the weights required, the sashes must be weighed fully glazed and calculated as follows:

$$\text{Top sash} \quad \frac{\text{Weight of sash} + 0.5\,\text{kg}}{2} = \text{mass for each weight}$$

$$\text{Bottom sash} \quad \frac{\text{Weight of sash} - 0.5\,\text{kg}}{2} = \text{mass for each weight}$$

The 0.5 kg difference with each sash is to assist when sliding and to keep them up or down in the closed position respectively.

3.2.13 Cords and chains

Figure 3.23 shows how cords and chains are attached to sash stiles, and how they support their weights. Cords are used mainly for domestic premises, whereas chains would be used on larger sashes, for example in hospitals, factories, etc.

A method of cording sashes is described and illustrated in Chapter 12.

3.2.14 Spiral balances

There are a number of proprietary spiral balances available on the market that eliminate the need for cords and weights. Solid frames

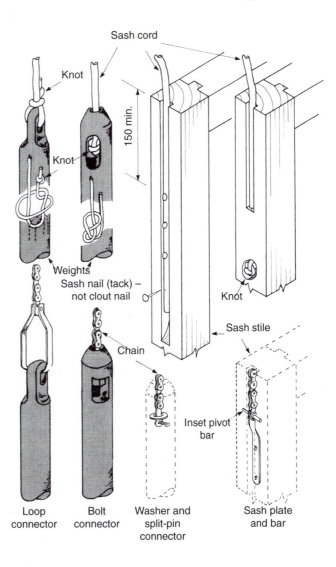

Fig. 3.23 *Methods of attaching cords and chains to sashes and sash weights*

Note: *Consult fixing instructions when fitting this type of spiral balance.*

are used to support the sashes (see section 3.2.10) to include a groove that is run either on the edge of the sashes or in the solid frame to accommodate the spiral spring (see Fig. 3.24a and b).

Figure 3.24 shows the sequence and procedures involved when fitting a 'balanceuk' spiral balance.

Regular cleaning and maintenance of the spiral rod with a light oil is recommended to improve the operating action of the spring. Other points to consider when fitting a spiral balance are not to:

• bend the spiral rod;
• forget to fit the limit stop;

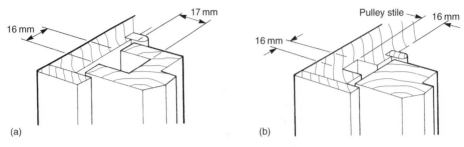

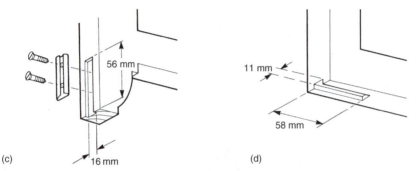

Spiral balance may be fitted into the sash stile (a) or pulley stile (b)

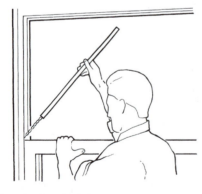

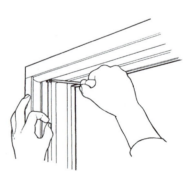

Top sash (c) and bottom sash (d) prepared to receive the foot of the spiral balance

(e) Inserting the spiral balance

(f) Securing the top of the spiral balance

Fig. 3.24 *Installing a 'balanceuk' sash balance*

- use the wrong spiral rod for the weight of sash;
- tension more than necessary;
- tension before glazing.

3.3 Metal casement in wood surround

With the possible exception of some types of louvred windows, windows made of steel and aluminium alloys are now generally fixed direct to masonry without a timber subframe, which in the case of aluminium-alloy windows was formerly thought necessary to avoid

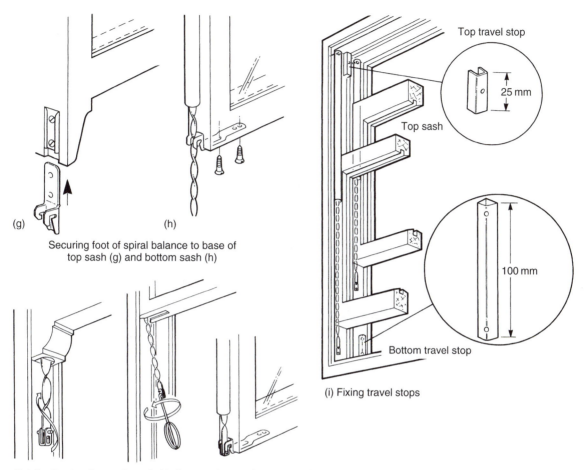

(g)   (h)

Securing foot of spiral balance to base of
top sash (g) and bottom sash (h)

(i) Fixing travel stops

(j) Adjusting tension on the spiral balance using tensioning hook (supplied)

Fig. 3.24   (*Continued*)

direct contact with the alkaline components of mortar and concrete (especially when damp). However, with the use of modern gaskets and sealants (chalking agents) the adverse reactions, which were previously expected, now seem to be avoided.

3.3.1 Metal casement windows

Because of the slenderness of the metal profiles of domestic types of window, the use of timber as a subframe is still popular. Wood species such as oak, sweet chestnut, western red cedar, and Douglas fir may, because of their acid content (particularly when damp), attack or react with steel. A subframe made from timber from these species should therefore either be sealed or have a gasket positioned between it and any metalwork.

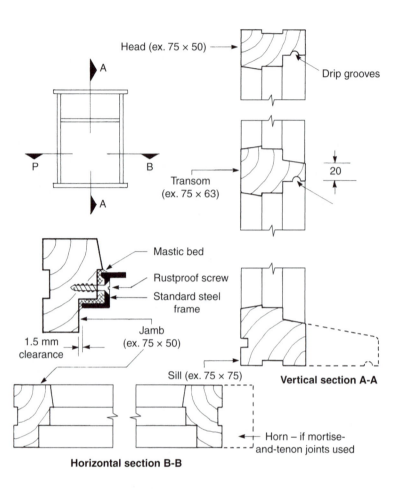

Fig. 3.25 *Sections through a timber subframe for a steel window*

Not only does a wood surround enhance the whole appearance of the window, it also facilitates fitting and removal if maintenance is required. Figure 3.25 shows typical vertical and horizontal sections through a wood surround suitable for a steel window. Its construction is similar to that described for a standard wood window frame. Aluminium-alloy windows may fit into a similar surround.

3.4 Glass louvred windows

With these windows, glass blades 152 mm to 170 mm wide are slotted into flanged clips at each end, which pivot about a frame with centres less than the glass width so that the blades overlap one another. The number of blades can vary from one to fifteen and – depending upon the design – any number of blades may be opened in unison. Opening mechanisms, either motorized or physical, operate on a lever principle

Fully louvred

Top-louvred, transom rail, and fixed light

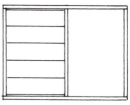
Louvred with fixed light

Fig. 3.26   *Glass louvred windows*

with coupled linkages. The operating handles (levers) are 'handed' (right-hand or left-hand) on the drive side of the frame, and the opposite side houses a following device.

The metal framework and flanged louvre clips may be made of galvanized steel or aluminium alloy. Flanged clips may also be made of plastics.

Figure 3.26 shows three possible louvre arrangements – configurations of both louvred and fixed lights are numerous.

Figure 3.27 shows how the 'Naco de luxe' louvred window is fitted and fixed.

1. The framework must be square (diagonals equal) and have a minimum width of 50 mm, at least 50 mm of which must be flat and square with its outer face.
2. Check for plumb and twist.
3. Position the operating channel plumb tight down on to the square sill.
4. Loosely fix the operating channel, then the non-operating channel in line with it.
5. Insert the PVC channel seal between the channel and the timber surround (see (a)), then tighten the screws.
6. Cut the head and sill weather-bars to fit between the channels.
7. Fix the weather-bars between the channels. The heel of the head weather-bar aligns with the inner face of the channels (see (b)). Use a non-setting mastic to seal the ends of the weather-bars and the underside of the sill weather-bar. Point the outside of the head weather-bar.
8. Insert glass blades as per the manufacturer's instructions. When the louvre window is closed, blades should touch each other throughout their length, and the upper and lower blades should touch the flexible PVC weather-tongue.

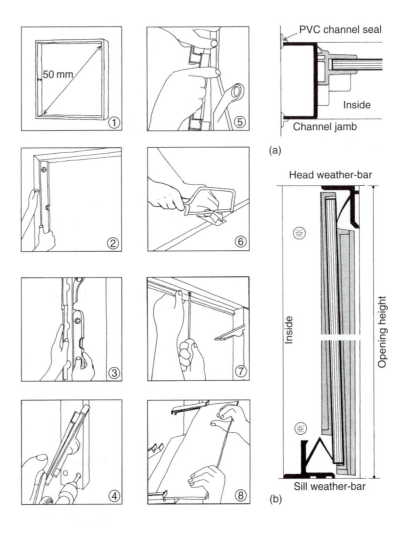

Fig. 3.27 *Fitting a 'naco de luxe' louvred window*

Figure 3.28 shows a fully louvred 'weatherbeta' works-assembled and sealed aluminium window.

Figure 3.29 shows how it could be fitted and fixed to a timber subframe.

3.5 PVCu (Polyvinyl chloride unplasticised) windows

With more emphasis being made to energy-efficient construction, timber may be substituted for the use of PVCu as a material for the construction of windows. However, we should mention here what is involved when manufacturing PVCu windows including possible pollutants in the air during manufacture, as opposed to timber, which is a renewable source of material.

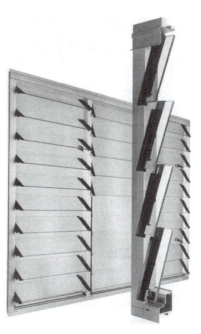

Fig. 3.28 *Fully louvred 'naco weatherbeta' sealed window*

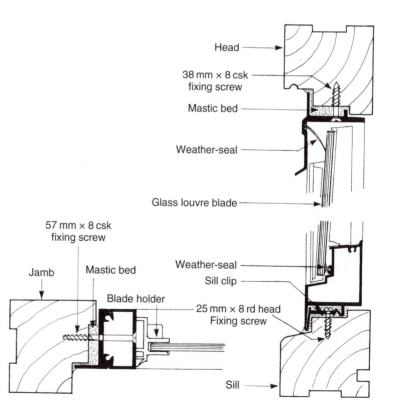

Fig. 3.29 *Louvred-window fixing details to a timber subframe*

Fig. 3.30 *Typical PVCu window*

PVCu window profile is a multi-chambered material, extruded usually with a bright white finish conforming to BS 7413: 2002 – other finishes are available.

Various window designs are available which are similar in style to the more traditional timber windows, as shown in Fig. 3.9, but incorporate, as standard, all the requirements necessary for PVCu to be energy efficient.

The example shown in Fig. 3.30 has sliding sashes supported on precision-made spring balances and includes 24 mm 'Low E' (Low Emissivity) glazing units, which is the term used to describe glass that has been treated with a surface coating which allows the sun's rays through, but reflects any heat back into the room.

Sashes, including those of timber construction, can be tilted inwards for maintenance, as well as allowing the glass to be cleaned (Fig. 3.31).

Note: *Approved Document F, Means of ventilation 2002, requires that all types of windows include a method of ventilation within the design of the window.*

3.6 Fixing window frames

Window frames are usually fixed into the wall structure in one of three ways:

(i) being built in as the wall is being built (Fig. 3.32);
(ii) inserting wall plugs and screws into a pre-formed opening (Fig. 3.33b);

Fig. 3.31 *Tilting the sashes for maintenance and cleaning*

(iii) fixing into pre-formed rebated reveals, for example double-hung cased windows Fig. 3.33d.

3.6.1 Built-in window frames

As shown in Fig. 3.32a, once the walls have reached sill height, window frames are positioned and held plumb by means of an anchored stay and rail at the head (hooked over the window head) ready to be walled in as brickwork progresses.

An arrangement of how a head and sill for a standard casement window may fit into the wall is shown in Fig. 3.32. Figure 3.32b shows how a vertical DPC (which must run the full height of the window) is attached to the back of the jambs with galvanized clout nails and is sandwiched between the outer wall leaf and the cavity closer. (Direct contact between an outer and inner wall leaf is not permitted without an intervening DPC – otherwise, moisture might be allowed to travel freely from the outer surface of porous brick or blockwork to the inner plasterwork. This is often a problem with older houses where vertical DPCs have been omitted.)

The methods of keying the sill and head horns, and providing mid-jamb anchorage into the wall, are similar to those used with door frames (see Chapter 4, Figs 4.5, 4.6).

All outside joints (small gaps) between the window frame and the wall must be sealed. Traditionally, these joints were made good and weathertight with a mixture of red mastic and boiled linseed oil, mixed to a stiff consistency, then applied with a narrow pointing trowel to form an angled fillet.

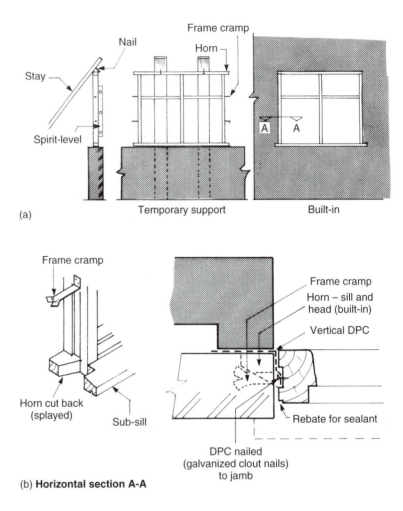

Fig. 3.32 *Fixing a built-in window frame*

(b) **Horizontal section A-A**

However, nowadays, modern sealants are available for most situations (see section 3.10); they are supplied in cartridge form and fit into a dispensing gun.

3.6.2 Non-built-in window frames

This method of installation is ideal for windows such as high-performance types and those made of polished hardwood that may be put at risk during the early stages of construction.

As shown in Fig. 3.33a, openings are formed by the bricklayer or mason with the aid of a removable template (profile) (see section 3.6.3) made to suit the window opening (size of window frame plus a fitting clearance of not less than 3 mm all round). To facilitate removal once the mortar has set, the profile should be allowed to reach a high moisture content before being built-in – on drying, subsequent shrinkage should then aid its removal. Planed (wrot) timber should be used for construction of the opening. Alternatively, for awkward and small openings, the

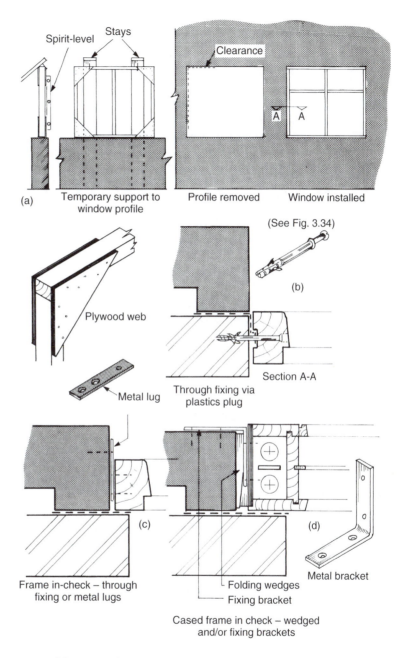

Fig. 3.33 *Fixing a non-built-in window frame*

top and bottom rails of the profile itself can be made detachable (see Chapter 4, Fig. 4.65) – the sides can then be easily withdrawn.

Propping and positioning profiles is the same as if window frames were being built-in. Unless they are required elsewhere, profiles can remain in position until it is safe to fit the windows. Covering the profiles with polythene will offer temporary weather protection to the inner structure.

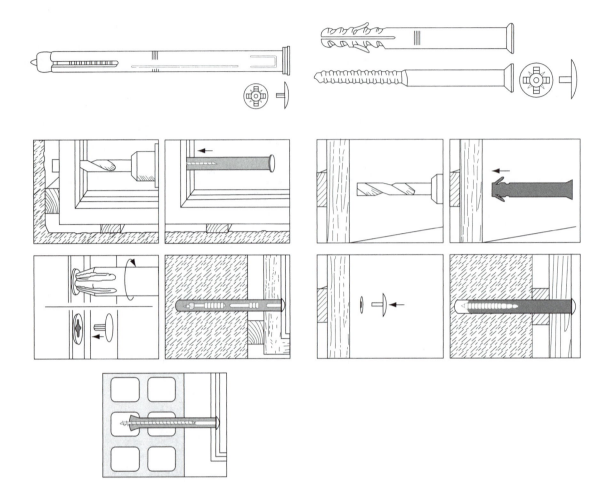

Fig. 3.34   *'Fischer' frame-fixing devices*

Cased frames of double-hung sash windows (Fig. 3.33d) are similarly positioned in check (see Fig. 3.33c) restricts window opening by forming a masonry rebate – mainly for two reasons: to slim down the appearance of the outer linings and, as with all rebated reveals, to provide a good weather-seal. The window is fixed in position with folding wedges driven between the reveals and the cased sides (in line with the head and sill). Skewed nails into plugs or steel angle brackets may also be used. Architraves are nailed to the inside linings (sides and head) and to the inside face wall via wood plugs or pallets to cover the main wedged fixings.

Joints between the framework and the outer wall are sealed as previously mentioned for built-in window frames.

3.6.3 Profiled openings

To aid the bricklayer when forming the openings for windows, timber templates of the window profile (Fig. 3.33) are constructed from planed timber, which are made in such a way as to be easily removed without damage to the surrounding brickwork, see section 3.6.2. These need to be made solid, square and braced as they may be used further elsewhere on the site.

3.6.4 Direct fixing

Figure 3.33b shows how jambs may be fixed with a flush reveal, using the type of fixings shown in Fig. 3.34. The plastics plug and sleeve are as one, and the whole assembly comes complete with a matching screw and provision for capping the screw head if the fixing is made through the jamb's rebate.

If the reveal is rebated, and thereby providing a positive outward window check as shown in Fig. 3.33c, inward movement may be checked by:

(a) plastics plugs and screws through the jamb;
(b) steel fixing plates, used as fixing lugs, screwed to plugged reveals.

3.7 Skylights (roof window)

As more and more people make use of their loft/attic space there has been an increase in loft conversions. Planning permission is required by the local authority if the area to be converted is to be 'habitable', i.e. extra bedroom or study, etc., which may also include 'en suite' facilities.

Note: *Consultation with a reputable building contractor together with an architect or building surveyor/designer who will draw up the plans and carry out all the necessary alterations is advisable.*

However, consideration must be given to means of access to the room by a staircase, which may mean the loss of valuable space from the floor below.

If the loft space is not to be used as a habitable area and a window is required just to provide light then planning permission is generally not required as long as the window is fitted in line with the pitch of the roof (Fig. 3.35) and does not effect the elevation. However, clarification from the local planning office must be sought before proceeding as required for a dormer window.

Note: *Provision must be made to prevent any loose items falling from the roof to the ground, whilst work is being undertaken.*

Many proprietary types of windows are available where the opening casement may be top-hung (Fig. 3.35) (recommended on lower pitched roofs) or pivot-hung (Fig. 3.36).

However, for a 'habital' room, the window must be of a certain design, by law, in order to act as a means of escape in an emergency – consult the window manufacturer.

Fitting of the window may be carried out from the inside of the roof space thus eliminating the necessity for a scaffold to be erected.

Fig. 3.35 *Top-hung skylights*

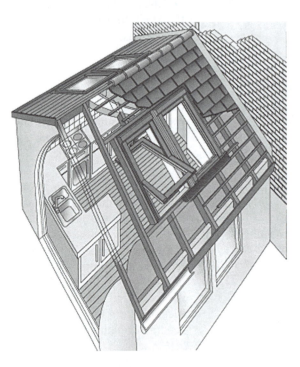

Fig. 3.36 *'Velux' pivot-hung skylight*

3.7.1 Installation

Before proceeding with any alterations to the roof, a risk assessment conforming to H&S regulations must be carried out before the work commences. This includes the area at ground level to be fenced-off with warning signs indicating workings above.

The installation procedure generally follows as:

1. install a solid floor inside the roof space;
2. remove the sarking felt (or underslating felt), if fitted;

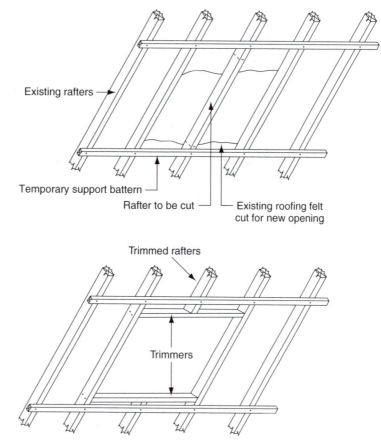

Existing rafters

Temporary support batten

Rafter to be cut

Existing roofing felt cut for new opening

Fig. 3.37 *Forming the opening, in a traditional pitched roof*

Trimmed rafters

Trimmers

Fig. 3.38 *Trimming the opening*

Note: *An existing truss rafter roof will need to be re-designed to accommodate any extra loading on the roof.*
Truss rafters and their components ***must never*** *be cut.*

3. remove the roof tiles to form an opening in the roof which is large enough to be trimmed out (see Volume 2, section 8.9 Roof openings) to suit the new window (truss rafters must not be cut);
4. cut the tile battens;
5. secure a suitable batten across the rafter/s to be cut, which is fixed to the remaining rafters that are not affected. This provides support to the rafter/s being cut and prevents damage to the roof and remaining tiles (Fig. 3.37);
6. remove the casement from the new window to make the frame lighter;
7. cut the rafter/s and trim to form the prescribed opening, in effect leaving an opening formed by two trimming rafters and two trimmers (Fig. 3.38) to the manufacturer's recommendations;
8. fix the frame to the opening using the brackets supplied. It is important that the frame is sitting square on the rafters;
9. fit the flashing and back gutter (supplied by the manufacturer) around the frame to make it watertight (Fig. 3.39);
10. cut and fit the apron lining into pre-formed grooves running around the inside edges of the frame, similar to fitting a window board (see Fig. 3.54c);

Fig. 3.39   *Fixing the flashing*

Fig. 3.40   *Rotating the sash for cleaning*

11. fit architraves around the opening to provide a suitable finish;
12. check all weathering before cutting and replacing roof tiles around the opening.

Finally, the casement is replaced back inside the frame and checked to make sure it is operating correctly.

Figure 3.40 shows how the sash can be rotated through 180° for maintenance and cleaning of the window.

A large window will mean a large opening in the roof's surface and greater loading on its members. In this case professional advice should be sought from either a structural engineer or an architect as to providing a means of accommodating any extra loading from the window. This could simply mean doubling-up the trimming rafters and trimmers, but no action should be taken until a full risk assessment is made by one of these professional bodies.

3.7.2 Dormer windows

When fitting this type of window to an existing property, it will require planning permission from the local authority as it affects the appearance or the shape of the original building.

In a new construction, the design of the dormer window varies and may include either a flat, pitched or curved roof (Fig. 3.41). Figure 3.42 shows a completed dormer window with a pitched roof and hipped end.

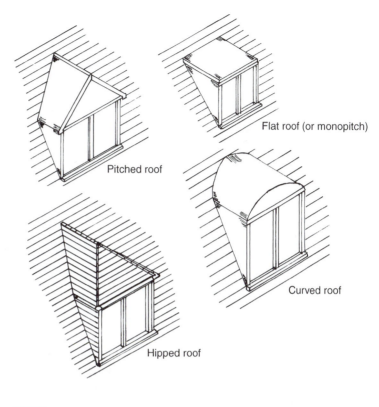

Flat roof (or monopitch)

Pitched roof

Curved roof

Hipped roof

Fig. 3.41 *Types of dormer windows*

Fig. 3.42 *Dormer window with hipped ended pitched roof*

3.7.3 Dormer construction

Because of the added load on the roof, dormer windows require additional support either via a 'double rafter' or a 'double floor' joist. This basically means two similar-sized timbers secured together, or a thicker size timber may be used, for example 75 mm.

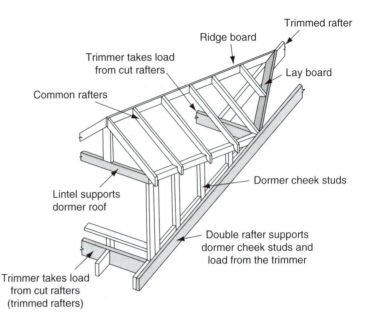

Fig. 3.43   *Dormer with pitched roof (gable end)*

Double rafters are included where the 'cheeks' (sides of the dormer) are fixed and supported by the roof structure (Fig. 3.43) (see also Volume 2, Chapter 8, Roofs of equal pitch, for terminology and construction). Construction of the dormer using the roof as support is as follows:

   (i)   position rafter/s in roof structure to form opening;
  (ii)   trim opening – to accommodate trimmed rafters;
 (iii)   cut and fix wall-plates and outer dormer cheek stud;
 (iv)   fix lintel to support the roof;
  (v)   cut and fix remaining cheek studs;
 (vi)   fix lay board, ridge support and ridge board;
(vii)   cut and fix common rafters.

Figure 3.43 shows details of a pitched roof dormer window with rafters that have been 'doubled-up' to provide the additional support. Trimmers are used to form the opening for the window as well as supporting a lay board for the flashing.

Dormers that are constructed with a flat roof (as shown in Fig. 3.44) require the cheeks of the dormer to be either supported by a double roof joist or a double floor joist.

Construction of the in-built dormer with a flat roof using double floor joists is as follows:

   (i)   position double floor joist where window is to be constructed;
  (ii)   cut and fix plate for vertical studs;
 (iii)   fix dormer cheek studs;

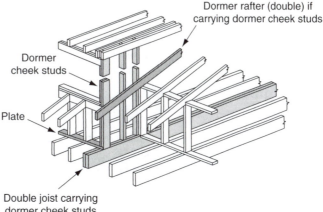

Dormer rafter (double) if carrying dormer cheek studs

Dormer cheek studs

Plate

Double joist carrying dormer cheek studs

Fig. 3.44 *Dormer with flat roof*

  (iv) position and fix outer dormer cheek stud;
  (v) cut and fix lintel and flat roof timbers to existing common rafters;
  (vi) trim out opening to support window;
(vii) cut and fix remaining cheek studs.

3.8 Glass and glazing

Window glass may be clear, patterned or wired, for use as follows:

(a) *Clear glass:* when produced by the 'float' process, this will provide undistorted through vision. Other clear glass production processes may produce glass with some image distortion.

(b) *Patterned glass:* this may be decorative and/or offer varying amounts of privacy, depending on its obscurity – the type and surface texture of the pattern are all-important factors.

(c) *Wired-glass:* this is used mainly where security is a factor or where a window contributes to the fire resistance of the building structure. The glass will crack when subjected to excessive heat but, because of the wire which is embedded within it, the whole pane will remain intact and planted within its rebate much longer than ordinary glass. Wired glass is available clear and patterned.

(d) *Fire protection:* this special type of glass provides a barrier to smoke, hot gasses, and flame as well as being an effective insulator against the heat of a blaze. Additionally, the glass will discolour, which will block out the fire to minimize panic and at the same time provide an indication to the fire services. This glass has an integrity (resistance to heat) of 30, 60 or 120 minutes depending on size and thickness.

(e) *Toughened:* this can be up to five times stronger that ordinary glass of the same thickness and is used in areas where security and safety are important. When broken it will break into relatively small pieces, thus eliminating the possibility of cuts to the skin. The glass is made stronger during the final stages by subjecting it to a heating and cooling process, which in turn sets up internal stresses.

(f) *Laminated:* this is manufactured with two or more layers of polyvinylbutyral (PVB) between the panes of glass to provide a reduction in noise and ensures that the glass, if broken, remains in place.

3.8.1 Size of glass

The thickness of glass used will depend on the surface area of the window pane and its exposure to wind.

In certain circumstances, where persons could be put at risk if accidental breakage occur (see BS 6262–4: 1994), glass sizes are restricted, for example low-level glazing, including doors and side window panels. Where such areas are to be glazed, special safety glasses are a necessary requirement (see Chapter 4, section 4.6.6, Critical glazed areas).

When designing the sash, where possible, particular attention should be made to the size of the glass panels in relation to the size of the sash (this may also include doors with glass panels, see Fig. 4.21).

Figure 3.45 shows, geometrically, a method to obtain the size of the glass pane. This is not only 'pleasing' to the eye, but is also in proportion to the window sash – known also as the 'Golden-ratio':

(i) set compass to line a–b (the width of the panel) and scribe an arc to give position c;
(ii) the compass is now set to b–c and an arc is scribed to give position d (the height of the panel).

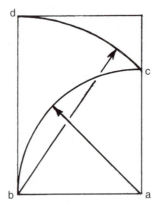

Fig. 3.45 *Obtaining the correct proportion for glass panels*

Traditionally putty is used for securing glass. It is a mixture of Linseed oil and inorganic fillers (BS 544: 1969) and applied using a putty knife onto primed timber. The putty becomes semi-rigid after time, by absorption into the timber and by oxidation – forming a skin.

3.8.2 Fixing glass

The rebate size will depend on the thickness of the glass or double-glazed unit and on the type of glazing material, for example putty or glazing beads or glazing compounds.

Note: *Linseed oil putty is not suitable when used with modern micro-porous paints and stains. (See BS 8000: part 7: 1990, Workmanship on site. Code of practice for glazing.)*

Rebates designed for Linseed oil putty fronting (pointing) should be at least 8 mm deep, unless they are for small frames. Non-setting compounds and tapes will need a rebate 10 mm to 12 mm deep. Rebates should be wide enough to receive 2 to 3 mm of back putty or compound, the thickness of the glass or unit, and the putty fronting or bead. It is important that rebates should be primed or sealed according to the type of putty or compound used.

Figure 3.46 shows a typical arrangement for a single-glazed window with putty-bed and fronting (pointing).

The window may be glazed in the following manner:

   (i)   Check that the glass is the right size for the opening, with 2 mm clearance all round (Fig. 3.46a).

  (ii)   Run a putty bed around the back of the rebate (Fig. 3.46b).

 (iii)   Position setting blocks (short pieces of sealed hardwood or PVC) in the bottom rebate (Figs 3.46c and 3.47). Location blocks (similar to setting blocks) may be required at the top and sides to prevent glass movement within a casement when it is opened. (See BS 8000: part 7: 1990, Workmanship on site. Code of practice for glazing.)

 (iv)   Position the glass in the frame and gently and evenly press around the edges until back putty squeezes out to leave at least a 2 mm backing (Fig. 3.46d).

  (v)   Hold the glass in position with glazier's sprigs (flat-cut nails) positioned at about 400 mm centres (Fig. 3.46e). Sprigs are tapped into place by sliding the head of a small hammer over the glass.

 (vi)   Cut back the back putty at an angle to produce a sight-line for the putty fronting (Fig. 3.46f).

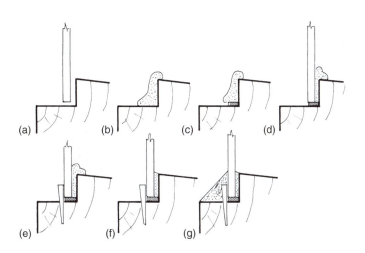

Fig. 3.46 *Single-glazed window with putty fronting (pointing)*

Note: *As soon as the putty has dried sufficiently it should be painted over, masking the gap left below the back sight-line. This will ensure a weather-seal between the glass, the putty fronting, and the framework.*

(vii) Run putty all the way round between the glass and the rebate. Then, starting at the top, cut the putty back to just under the back sight-line viewed through the glass, using a putty knife, a hacking knife (preferable), or an old chisel; then work down the sides and finally along the bottom in the same way (Fig. 3.46g). If the putty pulls away from the glass this can be remedied by gently running a thumb across or down the putty bevel.

Figures 3.47 and 3.48 shows the use of a front bead in place of putty fronting. Glass may be beaded externally, as shown, or internally in the case of extreme weather exposure and to provide greater security.

Beads may be nailed or screwed depending on the situation, with rust-proof screws sunk into screw cups to facilitate removal. Screws should be not more than 76 mm from the corners or 250 mm apart. Those parts of the bead which are to make contact with the bedding medium should be primed or suitably sealed to prevent absorption, as with rebates.

When non-setting compounds are used as a bedding, distance pieces of a non-absorbent resilient material, e.g. plasticized PVC, are set

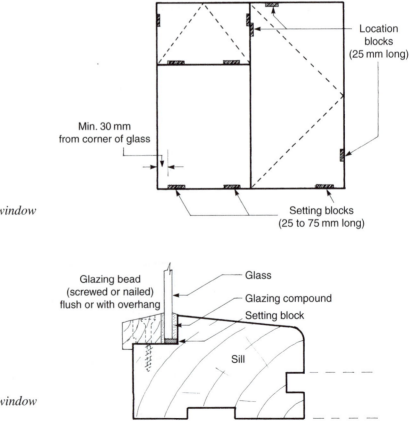

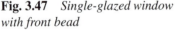
Fig. 3.47 *Single-glazed window with front bead*

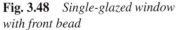
Fig. 3.48 *Single-glazed window with front bead*

between the rebate and the back of the glass to prevent the compound from becoming further compacted when the glass is subjected to high winds. They are usually spaced at under 300 mm centres.

3.8.3 Double and secondary glazing

There are three main reasons why windows are double or secondary glazed:

(i) To cut down the heat loss through single-glazed windows. Thermal insulation values increase as the air space between panes widens up to 20 mm – this optimum value may then be maintained until the gap reaches 200 mm.

(ii) To reduce the incidence of condensation on the glass by allowing the area within the inside surface of the glass to stay at a temperature above the dew point (the temperature at which the water vapour in the air condenses).

(iii) Double glazing can be designed to give good levels of noise (unwanted sound) reduction by:
   (a) providing good all-round airtight seals;
   (b) increasing the glass thickness;
   (c) providing an air space of between 100 mm and 200 mm between panes; using sound-absorbing material as a lining around the air space (secondary glazing – see Fig. 3.53);
   (d) ensuring a different thickness of glass for the outer and inner panes;
   (e) insulating the window from its surrounding wall structure with resilient material.

There are four different methods of providing double glazing:

(a) *Sealed double-glazed units* (Fig. 3.49): two panes of glass are held apart by a spacer, then filled with dry air or an inert gas and sealed. However, on more modern high-performance windows

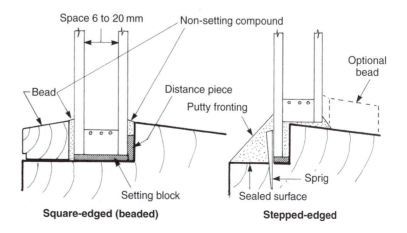

Fig. 3.49 *Fixing sealed double-glazed units*

Square-edged (beaded)   Stepped-edged

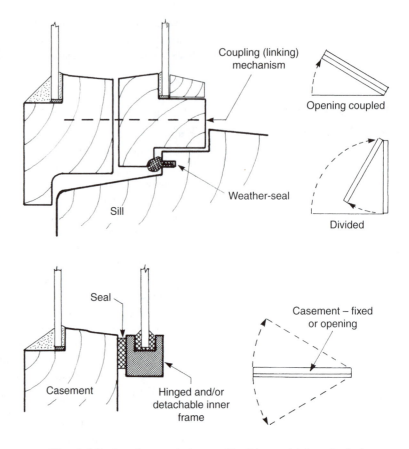

Fig. 3.50 *Coupled (linked) windows double glazing*

Fig. 3.51 *Secondary casements (sashes) double glazing*

(Fig. 3.16) the glass unit is usually 24 mm thick to include two 4 mm pieces of glass with a 16 mm gap.

Two types are available: 'square-edged' for wide rebates, and 'stepped-edged' for narrow rebates. The wider the space between the panes, the better the thermal insulation; but sound insulation values will, in the main, only be due to the double thickness of glass (increased mass).

(b) *Coupled windows (linked windows)* (Fig. 3.50): these consist of two casements hinged together, one of them also being hinged to the window frame. Normally, both open while linked together, but hinged separation is possible – usually for cleaning purposes.

(c) *Secondary casements (sashes)* (Fig. 3.51): these provide an add-on method of double glazing. A glazed framework of timber, metal, or plastics is attached to the inside of the existing window framework. These secondary casements may be hinged, made to slide, or static, but all should be detachable for cleaning purposes.

There are many proprietary systems on the market including the use of aluminium extruded channels. These are secured to the window reveal and support vertical or horizontal sliding glass units. Figure 3.52 gives an example of a proprietary horizontal sliding secondary glazing used in a window reveal.

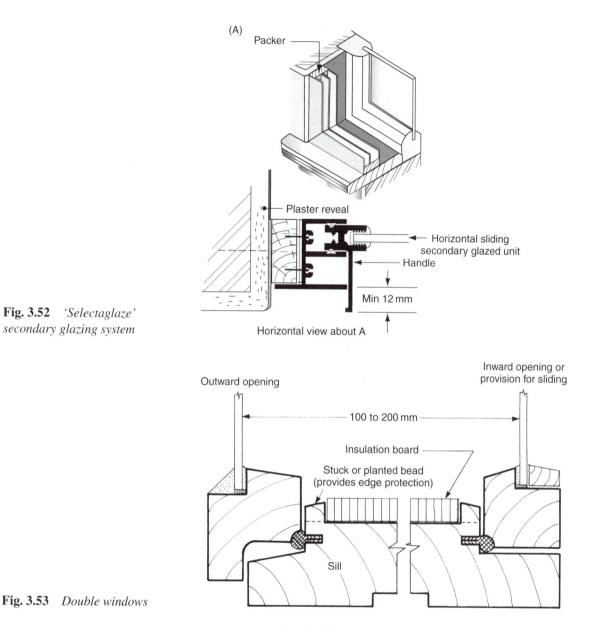

Fig. 3.52 *'Selectaglaze' secondary glazing system*

Fig. 3.53 *Double windows*

(d) *Double windows* (Fig. 3.53): these consists of two window frames in the same opening – set apart with a gap of between 100 mm and 200 mm to give maximum sound insulation as previously described.

3.9 Window boards (Fig. 3.54)

These can be made from solid timber, plywood, blockboard or MDF (medium density fibreboard). Their back edge is tongued into a pre-formed groove in the sill (Figs. 3.48 and 3.54a), while the board's width possibly (depending on the method of wall construction) spans a cavity.

The board is nailed or screwed down to the inner wall leaf via a series of packings and plugs. Packings are positioned at intervals to suit the board's thickness. The thickness of the packing can be determined by using a short end of window board, a wedge, and a short level as shown in Fig. 3.54b.

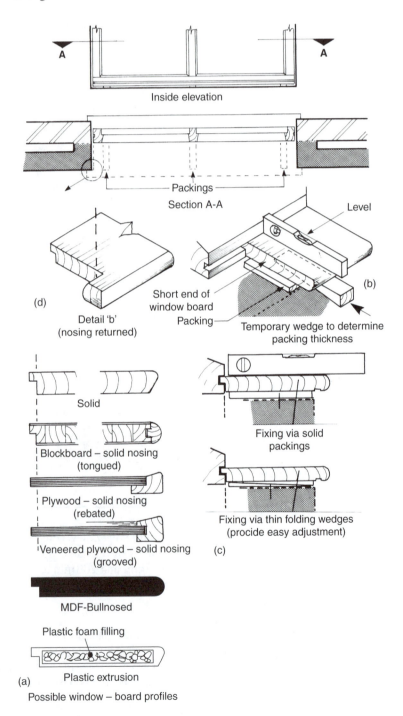

Inside elevation

Packings

Section A-A

Level

(d)

Detail 'b'
(nosing returned)

Short end of
window board

Packing

Temporary wedge to determine
packing thickness

(b)

Solid

Blockboard – solid nosing
(tongued)

Plywood – solid nosing
(rebated)

Veneered plywood – solid nosing
(grooved)

MDF-Bullnosed

Plastic foam filling

Plastic extrusion

Possible window – board profiles

(a)

Fixing via solid
packings

Fixing via thin folding wedges
(procide easy adjustment)

(c)

Fig. 3.54 *Window boards and
their fixing*

If the required thickness is non-standard, then folding wedges may be used – slide one upon the other until the correct thickness is obtained, then pin them together or nail them directly into the blockwork below (Fig. 3.54c).

Because window boards are generally fixed before plastering is carried out (first fix), there is no need for them to be a tight fit. It is usual to let the front edge overhang the plasterwork and extend by that amount beyond both reveals in which case, whatever front-edge finish or shape is used, it should be repeated and returned around each end as shown in Fig. 3.54d.

3.10 Sealants

To seal the gap between the frame and brickwork, it is necessary to provide some form of flexible sealant that will prevent the entrance of moisture into the building. The sealant also needs to be flexible enough to accommodate the movement of the building, as well as provide a continuous seal against the elements and match to its application (BS 6213: 2000).

Various sealants are available. These may be silicon or acrylic based and are available in cartridges and applied around the frame by the use of an applicator gun. The plastic nozzle may be cut to suit the diameter of the bead required. After application, the bead of sealant can be smoothed off by a damp rag, as well as any surplus before setting.

Expanded polyurethane foam is another type of sealant, which is applied between window or door frames and their surrounding openings after fixing, filling irregular gaps and improving insulation.

3.11 Protection on site

Windows should reach the site in a dry and semi-weather-protected condition having been previously treated with a good-quality priming paint and/or a water-repellent type of preservative at their place of manufacture.

3.11.1 Storage

Site storage should be such that windows are stacked flat or reared on edge – and not as shown in Fig. 3.55.

Figure 3.56 shows two examples of proper storage:

- flat on bearers with spacers to allow for ventilation;
- stacked vertically on edge (head to sill) on bearers with support to prevent falling over.

Fig. 3.55 *How not to store window frames on site*

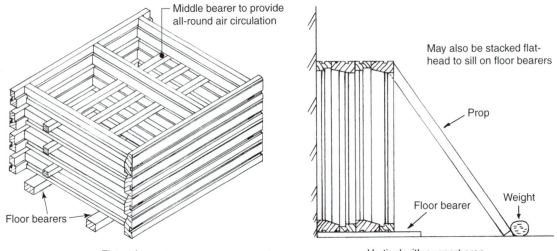

Fig. 3.56 *Examples of proper storage of window frames*

In either case windows must always be positioned out of twist, and preferably indoors. If they are to be stored in the open, stacks must be kept off the ground, free from rising damp. The top and sides of stacks should be protected from bad weather by a vented cover – air circulation within the stack is essential to counter any condensation which might otherwise form on the underside of the covering, particularly if plastic sheeting is being used. The stack should be assembled and covered in such a way that as windows are used from the stack, those that remain are left with adequate vented cover.

3.11.2 Preventative measures before and after installation

Before frames are installed, those parts (particularly end grain) that are to be enclosed within the wall structure, i.e. jambs, head, and sill should be fully protected by painting against any ingress of moisture,

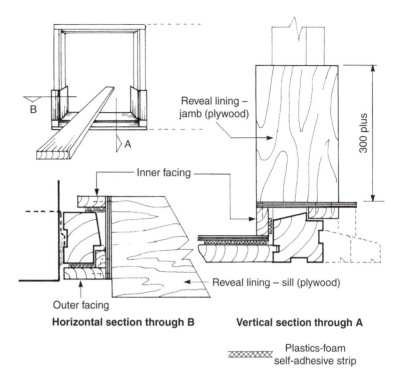

Fig. 3.57 *Sill and jamb protection against scuffing*

etc. If, for whatever reason, the remaining exposed woodwork cannot be fully protected against bad weather after installation, the whole framework should be clad with polythene or similar impervious material.

Where window openings are used as access points for lengths of timber, for example floorboard or skirting board, the exposed parts of the framework which may make contact (i.e. sills and the lower parts of jambs and mullions) should be temporarily boxed as shown in Fig. 3.57.

3.12 Window hardware (ironmongery)

Hardware has been divided into three sections:

- Traditional cased windows
- Modern and high performance windows
- Double-hung sash windows

3.12.1 Traditional (see Fig. 3.58)

(a) *Butt hinge:* used on traditional type sashes (Fig. 3.6) with each of the leaves of the hinge housed into the jamb and sash hanging stile. Positioning of the hinges should be just below the top rail and above the bottom rails so that end grain is avoided when letting in the hinges.
(b) *Stormproof hinge:* for use on stormproof sashes (Fig. 3.6). These are housed into the rebate of the sash hanging stile and into the jamb. Positioning is similar to that for the butt hinge.

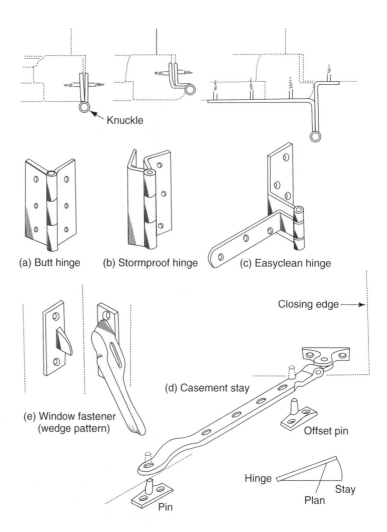

Fig. 3.58 *Casement window hardware (ironmongery)*

(c) *Easyclean hinge:* this type of hinge is surfaced fixed and not housed into the sash or jamb. Mainly used for windows above ground level, due to the fact that possible intruders could remove the hinges from outside. However, the advantages are that the hinge requires no housing-in, and when opened allows the sash to swing out enabling the outside of the glass to be cleaned.

(d) *Casement stay:* face fixed onto the bottom rail with two pins secured to the top edge of the sill, it allows the sash to be opened and secured in a number of set positions. When closing the sash, the casement stay is positioned behind the inner off-set pin and then hooked over the outer pin. By doing so, this will pull the sash tight into the rebate.

(e) *Window fastener:* secured to the face of the sash closing stile, with the 'striking plate' housed or surfaced fixed to the window jamb. Once the window is closed the handle locks into the wedge shape knuckle, thus locking and pulling the sash tight into the frame.

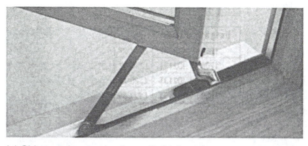

(a) Side opening mechanism with friction stay

Fig. 3.59 *Friction window stay*

(b) Fully reversible pivoting mechanism

3.12.2 Modern and high performance

Modern window hardware may have a dual function, for example a 'friction stay' shown in Fig. 3.59 that serves as both a hinge and a stay. Figure 3.59a shows the friction stay, in this case fitted to the base of the sash for concealment. When fully open the outside of the glass can easily be cleaned – ideal for windows above ground level.

The stay may also be fitted to the side of the sash (Fig. 3.59b) as top hung, again concealing the hinge and allowing for easy cleaning of the glass.

The 'Titon Axaflex' window stay (Fig. 3.60), however, doubles as a fastener. It holds the sash open at various positions and when closed locks it in place.

A variation on the two-pin casement stay (Fig. 3.58d) is the screw-down adjustable stay (Fig. 3.61), which allows the sash to be locked open at any position.

The window fastener shown in Fig. 3.62 includes a locking key for security and has two latching positions: one with the casement fully closed and the other with it partly open (13 mm gap at the closing edge) – useful as a night vent.

When the fastener shown in Fig. 3.63 is closed it activates two locking mechanisms: one from behind the handle and the other from the espagnolette bolt (rectangular steel bar), which is shot into the window head for greater security.

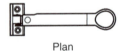

Plan

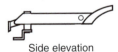

Side elevation

Closed position

Variable position

Fully open

Fig. 3.60 *'Titon Axaflex' window stay*

Fig. 3.61 *'TBKS' screw-down adjustable casement stay*

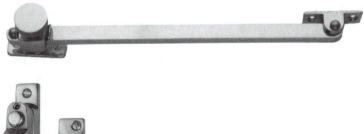

Security key

Fig. 3.62 *'TBKS' locking wedge fastener*

High-performance windows often include in their design a fastener like the one shown in Fig. 3.63 – its multi-point fastening device not only locates into the jamb, but also the head and sill for extra security.

Figure 3.64 shows the 'Trimvent®' (Trickle-vent) window ventilator, which is suitable for fitting through the window frame or sash top rail of both standard and high-performance windows.

Fig. 3.63 *Multi-point fastener*

Espagnolette fitted to closing edge of sash   Centre locking point

Trickle-vent in head of frame

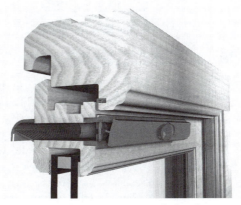

Fig. 3.64 *'Titon' Trimvent*®
window ventilator

Trickle-vent in top rail of sash

Fig. 3.65   *Sash pulley*

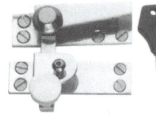

Lever fastener (locking)   Brighton sash fastener   Fitch fastener (locking)

Fig. 3.66   *Sash fasteners*

Fig. 3.67   *Sash lift*

3.12.3 Double hung sash windows

Hardware used on double-hung sash windows includes:

- Weights (Fig. 3.17); (see section 3.2.12);
- Chains/cords (Fig. 3.23); (see section 3.2.13);
- Spiral spring balances (Fig. 3.24); (see section 3.2.14);
- Sash pulleys (Fig. 3.65) – usually housed into the top of the pulley stiles (two each side) with the wheel made of either pressed Cast-iron, brass or nylon to accommodate the cord or chain;
- Sash fastener (Fig. 3.66) – shows three types available, each is secured to the sash meeting stiles and may be secured by a turn-screw or key;

- Sash lift (Fig. 3.67) – secured to the bottom rail it enables the bottom sash to be lifted to the required position.

References

BS 6262–4:1994, Glazing for buildings. Safety related to human impact.

BS 644: 2003, Timber windows.

BS 7413: 2002, Unplasticized polyvinyl chloride (PVC-U) profiles for windows and doors. Specification.

BS 6213: 2000, Selection of construction sealants. Guide.

BS 8000: 1990 part 7, Workmanship on site. Code of practice for glazing.

BFRC Window energy rating (WER) is now available rated between A and G on a BFRC window energy label, similar to home appliances ('A' being the most efficient).

CP 153: part 2: 1970 (Table 3).

Building Regulations Approved Document F: 2000.

Building Regulations Approved Document L1 and L2, Conservation of Fuel and Power: 2002.

Building Regulations Approved Document N, Glazing – safety in relation to impact, opening and cleaning: 2002.

4

Domestic doors, frames and linings

Doors, door frames and linings are an integral part of a domestic building and must provide a method of easy access or egress for any part of the dwelling without restrictions. This also applies to any outbuildings, which will require the same conditions.

There are many different door designs that may be straightforward and plain, or ones that include panels, glazed areas and even circular headed. However, whichever door is selected it must suit the dwelling and be aesthetic in design.

This chapter deals with a variety of doors and the methods used to support them, including the various types of ironmongery required for them to work efficiently.

4.1 Classification (single-leaf door patterns)

Doors may be divided into four main groups and may be used internally or externally according to their appearance and method of construction:
  (i) Unframed (matchboard) doors (Fig. 4.1):
      (a) ledged and battened;
      (b) ledged, braced, and battened.
 (ii) Framed doors (Fig. 4.1):
      (a) framed, ledged, and battened;
      (b) framed, ledged, braced, and battened;
      (c) panelled (solid and/or glazed).
(iii) Flush doors (Fig. 4.1):
      (a) plain;
      (b) panelled (vision panels).
 (iv) Special doors (Fig. 4.1):
      (a) shaped heads;
      (b) double margin;
      (c) stable door.

Doors may be further subdivided by their siting and/or function. For example, exterior doors must be designed to exclude the weather from the building and be strong enough to provide security – they are therefore generally thicker, heavier, and more substantial in construction than

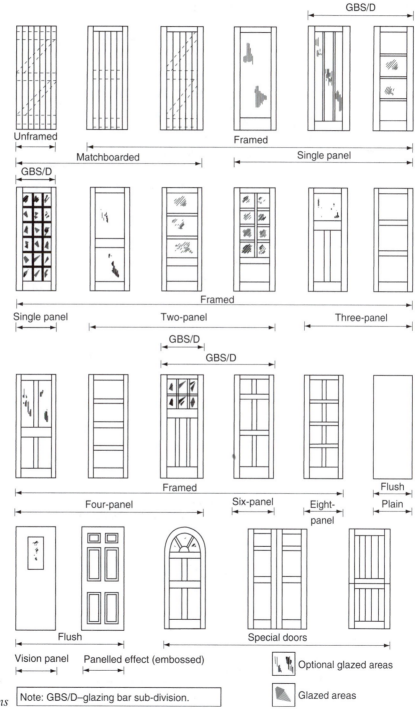

Fig. 4.1 *Examples of single-leaf door patterns*

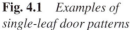

Note: GBS/D–glazing bar sub-division.

interior doors. Fire doors (section 4.10) provide fire resistance between compartments, etc. Louvred doors provide permanent ventilation.

4.1.1 Door types

Domestic doors are available in many sizes, which are indicated in Tables 4.1 to 4.3 – most are indicated in both imperial and metric sizes.

Table 4.1 Exterior door sizes

Height	Width	Thickness
Exterior metric single-leaf doors		
2000 mm	807 mm	44 mm
2040 mm	726 mm	44 mm
2040 mm	826 mm	44 mm
Exterior metric double-leaf doors		
2000 mm	1106 mm	44 mm
Exterior imperial single-leaf doors		
6′6″ (1981)	2′0″ (610)	1¾″ (44)
6′6″ (1981)	2′3″ (686)	1¾″ (44)
6′6″ (1981)	2′6″ (762)	1¾″ (44)
6′6″ (1981)	2′9″ (838)	1¾″ (44)
6′8″ (2032)	2′8″ (813)	1¾″ (44)
Exterior imperial double-leaf doors		
6′6″ (1981)	3′10″ (1168)	1¾″ (44)

Table 4.2 Internal door sizes

Height	Width	Thickness
Interior metric single-leaf doors		
2040 mm	626 mm	40 mm*
2040 mm	726 mm	40 mm*
2040 mm	826 mm	40 mm*
Interior imperial single-leaf doors		
6′6″ (1981)	2′0″ (610)	1⅜″ (35)
6′6″ (1981)	2′3″ (686)	1⅜″ (35)
6′6″ (1981)	2′6″ (762)	1⅜″ (35)
6′6″ (1981)	2′9″ (838)	1⅜″ (35)
Interior imperial double-leaf doors		
6′6″ (1981)	3′10″ (1168)	1⅜″ (35)
6′6″ (1981)	4′6″ (1372)	1⅜″ (35)

*Fire doors are 44 mm thick

Table 4.3 Mobility door sizes

Height	Width	Thickness
Mobility single-leaf doors		
2040	926	40*

*Fire doors are 44 mm thick

4.2 Ledged-braced and battened door (Fig. 4.2)

These types of doors (door leafs) are commonly used as exterior doors to outhouses, sheds, garages (for pedestrian access), and screens. Provided they are made and fixed correctly, they will withstand a lot of harsh treatment and remain serviceable for many years.

4.2.1 Construction

As can be seen from Fig. 4.2, the face of the door is made up of V or beaded tongue-and-groove matchboard, known as 'battens' (Fig. 4.2a).

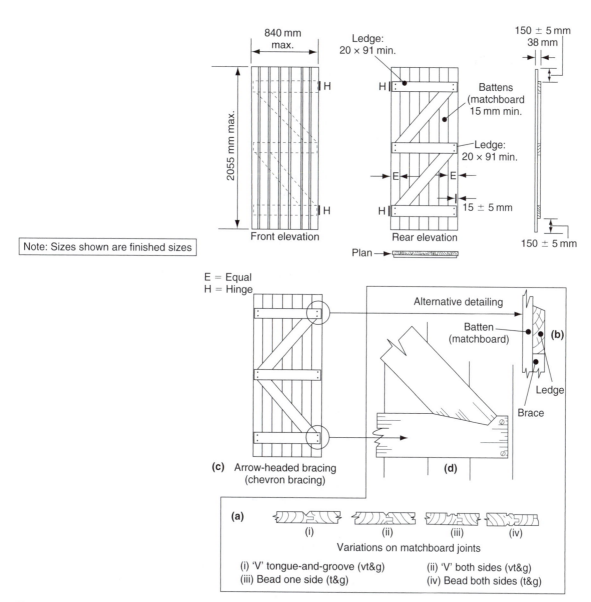

Fig. 4.2 *Ledged-braced and battened (matchboarded) door construction*

These are nailed or stapled to three horizontal members called 'ledges'. These ledges (if the door opens outwards the top edges should be bevelled, see Fig. 4.2b) are held square to the battens by similarly fixing two diagonal pieces of timber called 'braces' (Fig. 4.2d).

A less expensive version of this door is the ledged-and-battened door. As its name implies, it is built without a brace – its suitability therefore relies entirely upon its nailed construction, which under normal circumstances would prove inadequate. It can however, be used quite satisfactory in small openings or as a temporary door.

4.2.2 Door braces

The importance of a brace is illustrated in Fig. 4.3, where it will be seen that its effectiveness is controlled by its direction. If the door is to retain its shape, the brace *must* always point away and in an upwards direction from the hanging side (hinged side), otherwise the door could sag at the closing side (see Fig. 4.4) – hence the term 'sag bar' sometimes used when referring to a brace.

This means that this type of door should be handed (designed to hang either its left or right hand side). To avoid this, manufacturers either supply the braces loose to be fixed *in situ* (on site) to suit the handing, or fix them as a chevron pattern (Fig. 4.2c) – in this way only one of the braces is fully effective. To further increase the effectiveness of the brace, particularly in more heavier doors, the ends of the braces can be cut into the ledges using a bird's mouth joint (Fig. 4.2d).

Figure 4.4 shows how a ledged-and-battened (matchboard) door should not be fixed.

The making of a simple card or wood model (Fig. 4.5) should help clarify bracing principles:

1. Cut four pieces of stiff card, thin plywood, or hardboard, etc., 500 mm long by 50 mm wide to form the sides, top and bottom.
2. Join them together at the corners with a single nail, pin, or screw. This will allow each corner to pivot and produce a scissor movement.
3. Lay the frame flat. Move the corners until they are square (at 90° to each other).
4. Cut a piece of rigid material to fit between corners A and C – this will act as the brace.
5. While the brace is held in position take hold of side H. Lift the whole frame up and turn it until vertical. The brace should now be self-supporting, and therefore the frame will remain square. However, if the bracing piece has been omitted, or positioned into the opposite corners B and D, the frame would have collapsed. In

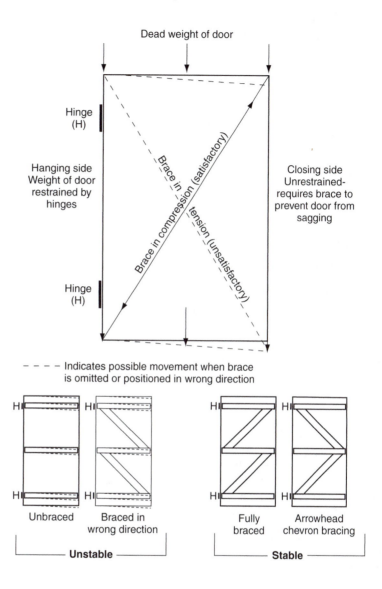

Fig. 4.3 *Bracing principle*

fact, corners B and D would have become wider apart (Figs 4.5a and b), so allowing the brace to fall out.

4.2.3 Assembly

Figure 4.6 shows some of the requirements specified by BS 459: 1988 for the construction of a ledged-and-braced battened door, but it is worth noting that traditionally such doors have been and are still in some cases made from heavier sectioned timber. They may also include other features such as those shown in Fig. 4.2, for example weathering to ledges (bevelling on their top edge) where the door is

Fig. 4.4 *Example of door sag (missing 'sag bar')*

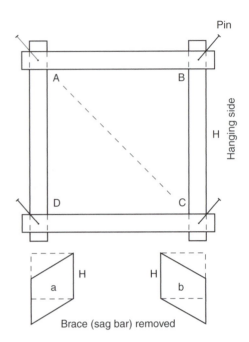

Fig. 4.5 *Model to simulate frame movement*

Note: *Widening gap at the top of the door (due to omission or wrongly positioned brace). Also the 'thumb latch' closing mechanism (see section 4.18, Hardware) should not have been used in this situation – the 'beam' is on the wrong side of the door face (see Fig. 4.77)*

subjected to weather from both sides (Fig. 4.2b), or extra brace restraint by housing braces into the ledges (Fig. 4.2d).

A typical order of assembly for a ledged-and-braced battened door is illustrated in Fig. 4.6.

Stage 1

(a) Cut the battens and ledges to length;
(b) Check that the ledges are not twisted (plane them out of twist if necessary);
(c) Paint or preserve all the tongues and grooves and the ledge faces that will come into contact with the battens;
(d) Rest the battens face down across the bearers, then lightly cramp the battens together;
(e) Double-screw each ledge to the cramped *edge* battens (Fig. 4.7).

Stage 2

(a) Cut the braces to fit between the ledges;
(b) Edge-nail the braces to the ledges (ensure that the door is kept square during this operation – if necessary use end stops).

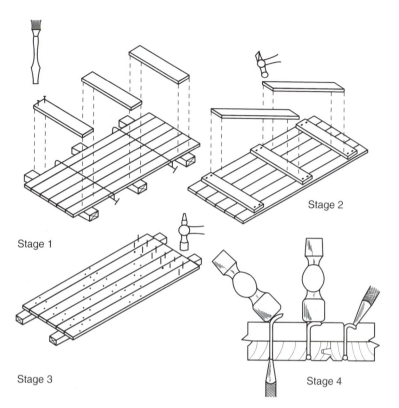

Stage 1

Stage 2

Stage 3

Stage 4

Fig. 4.6 *Stages of assembly*

Note: *Edge battens must be equal (E) and of sufficient width to allow for fixing. This is achieved by placing either a batten or joint to the centre of each ledge.*

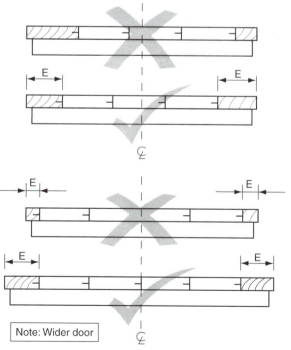

Note: Wider door

Fig. 4.7 *Positioning edge battens*

Table 4.4 Sequence of operations for ledged braced and battened door construction

Process	Item	Member	Operations	Machines	Powered hand tools	Hand tools	Ironmongery	Remarks
Preparing timber	Door	Ledges, battens, braces	Cut to length, ripping, deeping	Cross-cut circular saw bench	–	–	–	Cut to nominal size (Ex)
			Planing	Hand-feed planer/surfacer, panel thicknesser	Planer	Hand plane	–	Finished size – width and thickness
		Battens	Form V, tongue-and-groove	Spindle moulder	Router	Rebate plane Plough plane	–	Worked by machine or hand
Assembly	Door	Battens, ledges, braces	Fix battens to ledges	–	Nail gun, power driver	Hammer	Nails/staples	Paint or preserve T & Gs – leave to dry before assembly
			Cut bevel and fix	–	Chop saw	Panel saw, hammer	–	Boards held face to face during treatment
			Punch and clench nails	–	–	Hammer, nail punch	–	
Finish	Door		*Sanding	Belt sander	Belt, Orbital or random	Cork sanding block, glass or garnet paper	–	Finish depends on grade of abrasive usually before assembly
	Door		Knotting Painting	–	–	Paint brush	–	Shellac knotting primer paint

* This depends on the type of finish. For varnishing/polish finish a belt sander would be used, with the grain. For a paint finish an orbital/random sander is used to prepare the surface and help with the adhesion of the paint to the wood

Stage 3

(a) Turn the door over on to bearers which have been placed length-wise;

(b) Double-nail each batten to the ledges and braces taking care to avoid the bearers as after the nails have been punched below the surface they will protrude 10 mm through the door.

Stage 4

(a) Turn the door back onto its face, where the protruding point of each nail will be visible;

(b) Clench each nail by bending it over in the direction of the grain, then punch the clench below the surface of the wood.

An overall sequence of operations can be listed in chart form as an easy means of reference. Table 4.4 shows how the door's construction can be tabulated.

4.3 Framed-ledged-braced and battened doors

The only difference between this type of exterior door and the ledged-braced-and-battened door is the addition of a frame around the matchboarding.

4.3.1 Construction

This door is constructed of all solid softwood and is designed to withstand a reasonable amount of rough treatment. It is quite often used in and around farm and industrial buildings.

The frame, as shown in Fig. 4.8, has two stiles and a top rail, which are of the same thickness. The bottom rail and the middle (lock) rail are thinner, to allow for the batten thickness plus 2 mm. All joints are either: through mortised-and-tensioned (the bottom and middle rails having bare-faced tenons) with proportions as illustrated; glued, wedged, and pinned with hard wood or metal (star) dowels; or dowel-jointed as described for panelled doors (section 4.5).

Braces should be a close fit between the stiles and the rails, be butted or tongued, and be positioned to slope upwards and away from the hanging side of the door (see Fig. 4.2).

Battens (matchboard) should be tongued-and-grooved with V or bead joints and tongued or rebated into the stiles and top rail, then nailed or stapled (with a pneumatic nail gun) to all framing members.

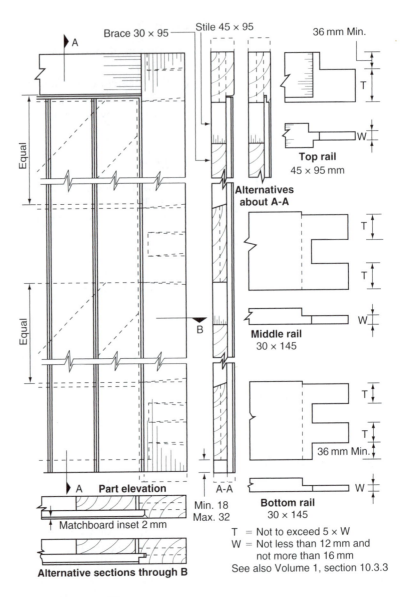

Brace 30 × 95

Stile 45 × 95

36 mm Min.

T

W

Top rail
45 × 95 mm

Equal

Alternatives about A-A

T

T

W

Middle rail
30 × 145

B

Equal

T

T

36 mm Min.

W

A ▶ **Part elevation**

Bottom rail
30 × 145

A-A

Min. 18
Max. 32

Matchboard inset 2 mm

T = Not to exceed 5 × W
W = Not less than 12 mm and
not more than 16 mm
See also Volume 1, section 10.3.3

Fig. 4.8 *Framed-ledged-braced and battened door*

Alternative sections through B

All tongues, grooves, and rebates should be painted or treated with wood preservative before assembly. (See Volume 1, Chapter 3, taking into account the safe use and storage of wood preservatives.)

4.4 Stable door

This type of door was originally used on farm stable buildings when housing horses or ponies, but has become more popular recently in domestic buildings as a back door into the garden from the kitchen or utility room (Fig. 4.9). It is divided into two halves, which allows the top half to be open while the bottom half can be securely shut. When

Fig. 4.9 *A stable door in use*

originally used on farm buildings, it allowed fresh air into the stable but contained the animal within.

4.4.1 Construction

Figure 4.10 shows the construction of the door, which is very similar to that of the framed-ledged-braced-and-battened door. However, the main difference is the middle rail, which has been divided by rebates across the full width of the door. This allows the two leaves, when closed, to form a solid meeting rail that prevents any draughts from entering the building. Additionally, weathering is included on the door to prevent any possibility of moisture from entering.

Traditionally, matchboarding would be used for the exterior, but some manufacturers prefer the use of manufactured boards (for example exterior MDF) with simulated grooves on the surface. Additionally, a glazed area may be provided in the top half of the door.

4.5 Panelled doors

These consist of a softwood or hardwood outer framework housing one or more panels of solid wood, manufactured boards, glass, or a combination thereof.

Figure 4.11 identifies the various members used in the construction of a four-panelled door. Panel style (profiles) may vary as shown in Fig. 4.14.

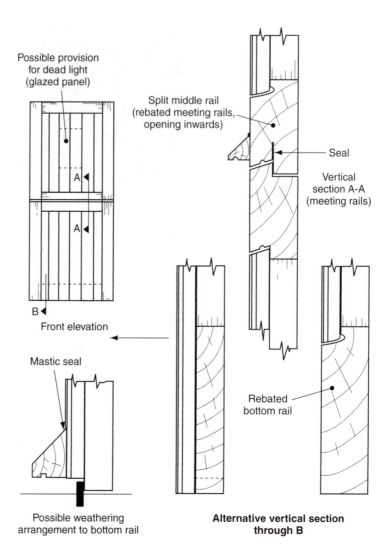

Fig. 4.10 *Stable door construction*

4.5.1 Construction

The frame includes two stiles, with two or more rails (top rail, bottom rail, middle or lock rail, glazing bars, and/or intermediate/frieze rails) and in some cases intermediate vertical members (muntins and/or glazing bars).

Joints between members may be mortised-and-tenoned or dowel-jointed.

Glazing bars are mortised-and-tenoned (see Chapter 3, Figs 3.22c to f for double-hung sashes).

4.5.1.1 Mortise-and-tenon joints (Fig. 4.13a)

The top, bottom, and at least one more rail (except with single-panel doors) should be through-jointed. Stub tenons into stopped mortise

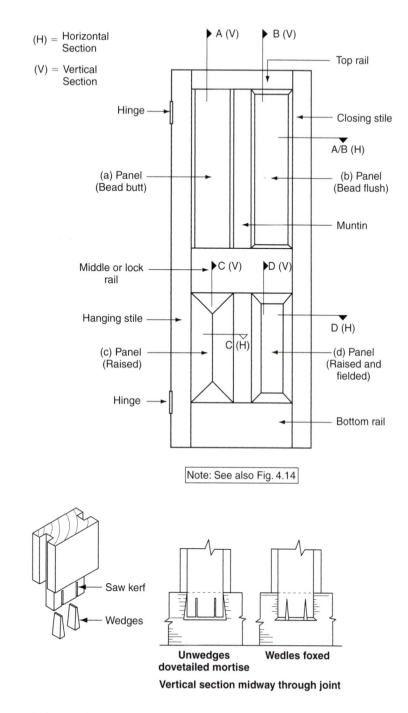

(H) = Horizontal Section

(V) = Vertical Section

A (V)　　B (V)

Top rail

Hinge

Closing stile

A/B (H)

(a) Panel (Bead butt)

(b) Panel (Bead flush)

Muntin

Middle or lock rail

C (V)　　D (V)

Hanging stile

D (H)

(c) Panel (Raised)

C (H)

(d) Panel (Raised and fielded)

Hinge

Bottom rail

Note: See also Fig. 4.14

Fig. 4.11 *Traditional four-panelled door with alternative panel arrangements*

Saw kerf

Wedges

Unwedges dovetailed mortise

Wedles foxed

Vertical section midway through joint

Fig. 4.12 *Fox-wedged stub tenons*

holes can be used to join any other rails to stiles or muntins to rails – fox wedging may be employed as shown in Fig. 4.12.

Tenon thickness should in nearly all cases be as near to one-third the thickness of the members being joined as is practicable (depending

Note: *Hand-made doors may be part-scribes as shown in Chapter 3, Fig. 3.22a*

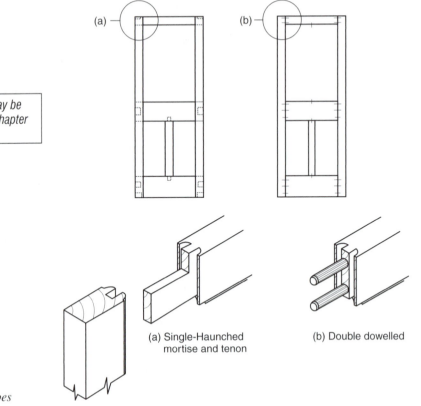

(a) Single-Haunched mortise and tenon

(b) Double dowelled

Fig. 4.13   *Through-scribes*

on chisel size). The tenon width should not exceed five times its thickness. Haunches should be as deep as the tenon is thick and never less than 10 mm. Tenons should be clear of the top and bottom of the door by a distance not less than 36 mm – see Fig. 4.8. All joints should be glued and wedged as necessary.

Where moulding has been worked (stuck) on to the inside edge or edges of the door framework, rail ends should be through-scribed as shown in Fig. 4.13a.

4.5.1.2 Dowelled joints (Fig. 4.13b)

Dowels should be made of hardwood or of the same species as the members being jointed – not less than 16 mm in diameter and 125 mm in length and at centres of not more than 55 mm, with no less than the number of dowels per joint as shown in Table 4.5.

Dowels to bottom rails should be not less than 45 mm up from the bottom of the door. All dowels must be fluted and joints glued. Joints to moulded sections should be through-scribed, with tongues to help stiffen the joints.

Table 4.5 Minimum number of dowels between interconnecting members

Joints between members	Minimum number of dowels
Stiles and middle/lock rail	3
Stiles and top rail	2
Intermediate members	1*

*Provision must be made to prevent turning

4.5.2 Panels

Figure 4.14 relates to the door in Fig. 4.11 and shows how manufactured boards and solid wooden panels may be set within the door framework. All panels should have a certain amount of clearance all round, within the groove in which they fit, to allow for moisture movement; plywood should be given a minimum of 2 mm all round, solid wood 3 mm. Veneered panels may be included to match the timber of the door to minimize movement; these are less expensive than solid timber panels.

Edges between the frame and panel may be left square but are usually finished by one of the methods shown in Fig. 4.15.

(a) *Stuck moulding:* in this case a moulding (ovolo) has been worked on both sides of the groove which houses a plywood panel – *panels must be built into the door as it is being assembled.*
(b) *Planted moulding:* the groove in this case has to be formed by nailing a planted mould (bead) onto both sides of the glass or plywood panel – *these panels are fixed after the door is assembled.*
(c) *Stuck and planted moulding:* the rebate is formed with a stuck moulded ovolo, onto which the plywood or glass panel is checked and then held in place with a planted ovolo moulded bead – *again the panels are fixed after the door has been assembled.*
(d) *Bolection moulding:* panels are built into the door framework as it is assembled – mouldings surround the panel on both faces of the door. On one face the bed moulding is slightly recessed, on the other face the bolection moulding is proud of the door. Bolection moulds are rebated on one edge to allow them to stand proud above both the panel and surrounding framework – they are held in place by screws set through the panel. The screw heads are masked by the bed moulding which is nailed to the panel framework.

Mouldings (beads) should not restrict any moisture movement of a panel – particularly a solid wooden panel. Small mouldings can be attached to the frame. Larger cover mouldings (bolection mouldings) can be fixed to the panel.

Glass (borrowed light – see section 4.6.5) as a door panel is very popular in both exterior and interior situations. Glass size – particularly of

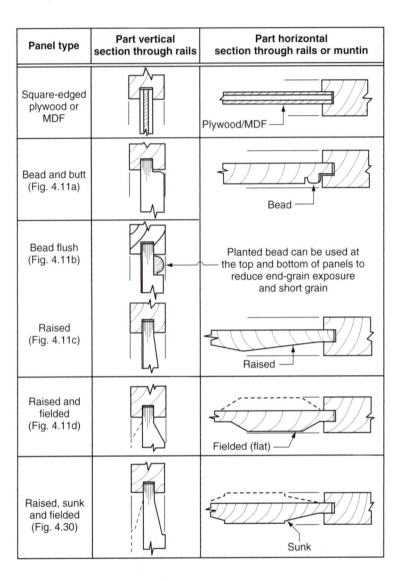

Panel type	Part vertical section through rails	Part horizontal section through rails or muntin
Square-edged plywood or MDF		Plywood/MDF
Bead and butt (Fig. 4.11a)		Bead
Bead flush (Fig. 4.11b)		Planted bead can be used at the top and bottom of panels to reduce end-grain exposure and short grain
Raised (Fig. 4.11c)		Raised
Raised and fielded (Fig. 4.11d)		Fielded (flat)
Raised, sunk and fielded (Fig. 4.30)		Sunk

Fig. 4.14 *Vertical and horizontal sections through different panel types (see Fig. 4.11)*

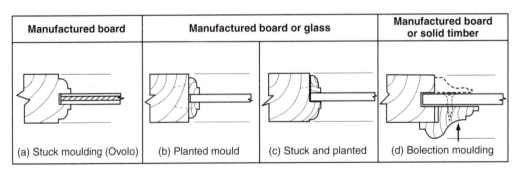

Manufactured board	Manufactured board or glass		Manufactured board or solid timber
(a) Stuck moulding (Ovolo)	(b) Planted mould	(c) Stuck and planted	(d) Bolection moulding

Fig. 4.15 *Detailing between frame edge and panel*

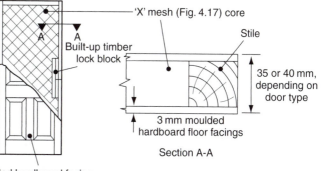

Fig. 4.16 *Panel effect moulded door*

Moulded hardboard facing in one piece over both faces

an unbroken (i.e. without glazing bars) area within a door – is very important because of the high risk of injury if breakage occurs. Recommendations for such situations can be found within BS 6262–4: 1994, Glazing for buildings. Safety related to human impact – see also Chapter 3, section 3.8, Glass and glazing.

4.5.3 Moulded panel doors

Figure 4.16 shows a type of panelled door made up using a timber framework containing a cellular core that has been clad both sides with either hardboard or plastics (PVCu).

Both faces of the door have been moulded to give the appearance of a traditional panelled door, which can include an aperture for glazing. Generally, moulded doors with plastics facings are used in external situations, whereas internal doors are faced with hardboard that has been pre-treated with a primed textured wood grain finish suitable to receive a paint or wood stain.

4.6 Flush doors

Except for glazed and louvred apertures, the whole outer surface of a flush door should be perfectly flat.

Both interior and exterior qualities of doors can be constructed – general construction and the type of adhesive used in the manufacture of the facing material being the main differences. Other factors include door thickness, rigidity, and provision for fixing hardware, etc.

4.6.1 Construction (Fig. 4.17)

With the exception of the solid (laminated timber) door, some form of framework will be required to retain the core of the door and/or provide a fixing for rails, etc.

Top, bottom, and intermediate rails should be notched or bored (Fig. 4.18) to allow movement of air through the door to help prevent

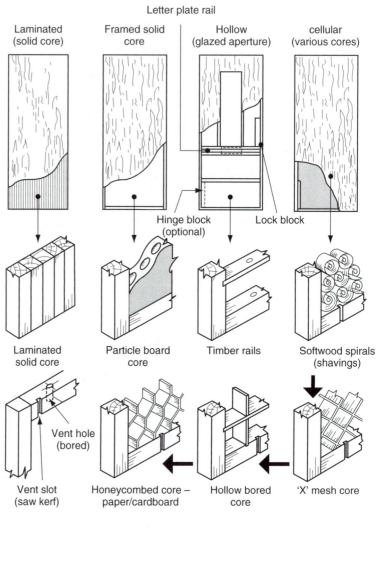

Letter plate rail

Laminated (solid core)

Framed solid core

Hollow (glazed aperture)

cellular (various cores)

Hinge block (optional)

Lock block

Laminated solid core

Particle board core

Timber rails

Softwood spirals (shavings)

Vent hole (bored)

Vent slot (saw kerf)

Honeycombed core – paper/cardboard

Hollow bored core

'X' mesh core

Fig. 4.17 *Framework for flush doors – together with their core*

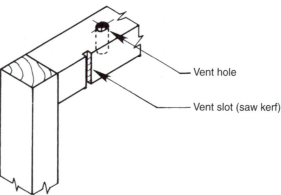

Vent hole

Vent slot (saw kerf)

Fig. 4.18 *Venting top, intermediate, and bottom rails*

distortion (air becoming trapped within the core of the door) at the assembly stage and so that, in the case of interior doors, air within can be at equilibrium with that which surrounds it.

The framework is assembled in a very simple fashion. Joints may be mortised-and-tenoned, but are more likely to be tongued-and-grooved or butted and stapled together. With cellular doors, the main strength of the door is derived from the stressed skin linings (facings), which clad both sides of the frame and its core.

4.6.2 Cores (Fig. 4.17)

Solid cores include laminated timber and sheet particle board (chipboard or flax board). Hollow cores include rails tenoned into mortised stiles or tongued into grooved stiles. The cellular boxes are formed using cross-halved timber or hardboard, paper/card strips, or softwood spirals formed by shaving short blocks of wood of uniform thickness.

4.6.3 Facings (Fig. 4.19)

Facings are generally plywood or hardboard, often with a decorative face veneer. Exterior-door facings should be of weather-resistant material, for example exterior-grade plywood or oil-tempered hardboard. Facings are glued under pressure to the framework and core.

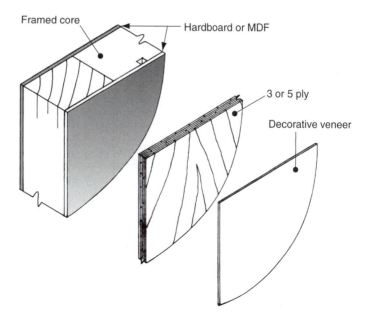

Fig. 4.19 *Door facings*

4.6.4 Lippings

These are thin laths glued to both side edges of the door to cover the edges of the facings. A species similar to the facings should be used on polished doors.

Figure 4.20 shows various methods of door edge treatment.

4.6.5 Glazed apertures (see also section 4.6.6 and Approved Document N)

These enable light to be transmitted ('borrowed light') from well lit areas and serve as vision panels to help avoid collisions. Size and shape can vary, for example square, rectangular, and circular are all possible, but for the door to retain rigidity, edge margins should be kept to a minimum distance of 127 mm at the top and sides and 200 mm up from the bottom of door, as shown in Fig. 4.21. If the core is non-solid, rails and blocking pieces must be provided to stiffen the door and provide a fixing for glazing beads.

Figure 4.22 shows two types of beading suitable for interior and exterior situations. Fixing with brass screws and countersunk cups is preferred to nailing. Beading that is flush with the face of the door is not

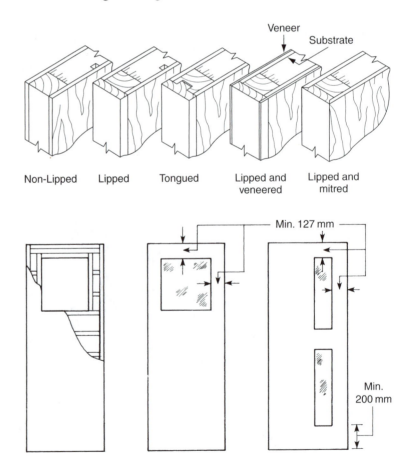

Fig. 4.20 *Flush-door edge treatment*

Non-Lipped   Lipped   Tongued   Lipped and veneered   Lipped and mitred

Veneer
Substrate

Min. 127 mm

Min. 200 mm

Fig. 4.21 *Margins to glazed apertures (vision panels)*

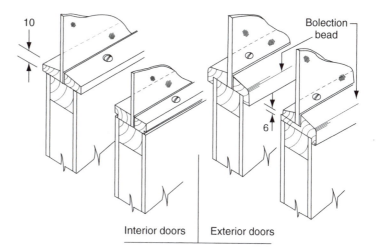

Fig. 4.22   *Glazing beads*

recommended, as this will cause the surface finish (paint or varnish) to crack along the joint at the edge of the door due to the timber movement of the beading.

Before exterior doors are beaded, the aperture edges should be sealed – waterproof tapes can be used.

4.6.6 'Critical' glazed areas (doors and screens) (Fig. 4.23)

Glass as a material when used in doors can be very dangerous when broken, especially at positions where there is a possibility that this may occur – either accidentally or on purpose (vandalism). For this reason glass must be strong enough to withstand human impact (see Building Regulations Approved Document N: 2000, BS 6206: 1991, and Chapter 3, section 3.8).

In Fig. 4.23 the areas indicated are 'critical' areas in terms of safety and if broken must disintegrate into small particles that are not sharp or pointed. This also applies to windows.

4.7 Louvred doors

In most cases, the purpose of a louvre sited within a door is to accommodate the need for permanent ventilation through it. Figure 4.24 shows how this is achieved, while at the same time offering privacy and, in the case of an exterior door (depending on blade pitch and projection), weather protection. Figure 4.25 shows how louvres may be sited to suit different types of door.

4.7.1 Construction

Figure 4.26 shows details of how a lightweight interior louvred wardrobe or airing cupboard door could be constructed.

Note: *Glazed areas should not be placed at the base of a staircase where a person may trip and fall*

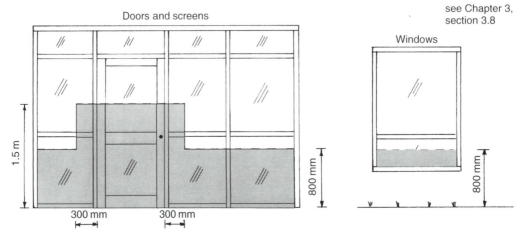

Fig. 4.23   *'Critical' glazed areas in doors, screens and windows*

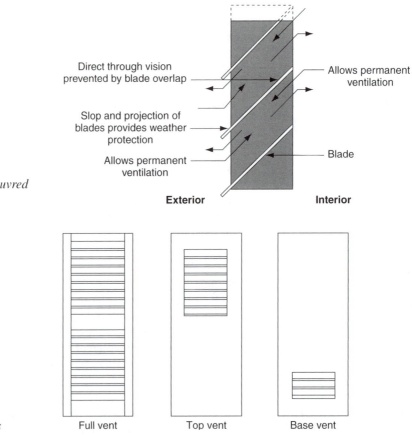

Fig. 4.24   *Function of louvred openings*

Fig. 4.25   *Louvred doors*

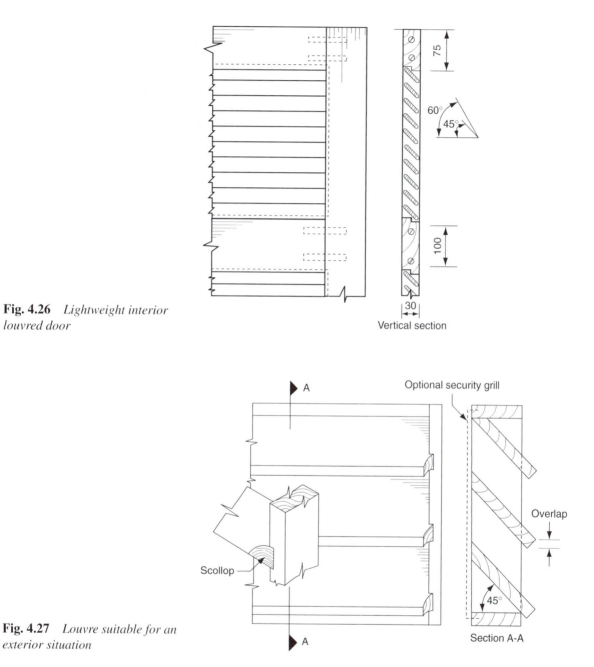

Fig. 4.26 *Lightweight interior louvred door*

Fig. 4.27 *Louvre suitable for an exterior situation*

Louvres in exterior doors usually have to provide security. They are therefore more sturdy in construction – as shown in Fig. 4.27. If the louvred area is such that when the louvres are removed this leaves an aperture large enough for a person to pass through, then this becomes a security risk. In this case, a secondary barrier in the form of a steel grille (or similar) should be considered.

4.8 Double margin doors (Fig. 4.28)

This type of door is used where an opening is too wide to accommodate a normal size door and not wide enough for two separate doors, which would otherwise look out of proportion. For this reason a double margin door is used to give the impression that there are two separate doors, but in fact there is only one door.

4.8.1 Construction

Generally, the door incorporates panels surrounded by a frame to include stiles, meeting stiles, top rails, frieze rails, middle rails and bottom rails. Wedged mortise-and-tenon joints are used throughout as well as folding wedges (Fig. 4.29) and a loose tongue to secure the meeting stiles (Fig. 4.30).

The door is assembled first by jointing together the meeting stiles using glue and three sets of folding wedges, which need to be cut back below the depth of the grove running around the internal edges of the frame. This prevents the ends of the wedges protruding through into the groove if any shrinkage takes place in the timber, which could in turn damage the panels.

Note: *See Figs 4.29 and 4.30 for details*

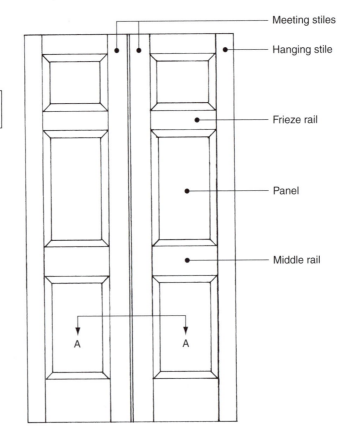

— Meeting stiles
— Hanging stile
— Frieze rail
— Panel
— Middle rail

Fig. 4.28 *Double margin door*

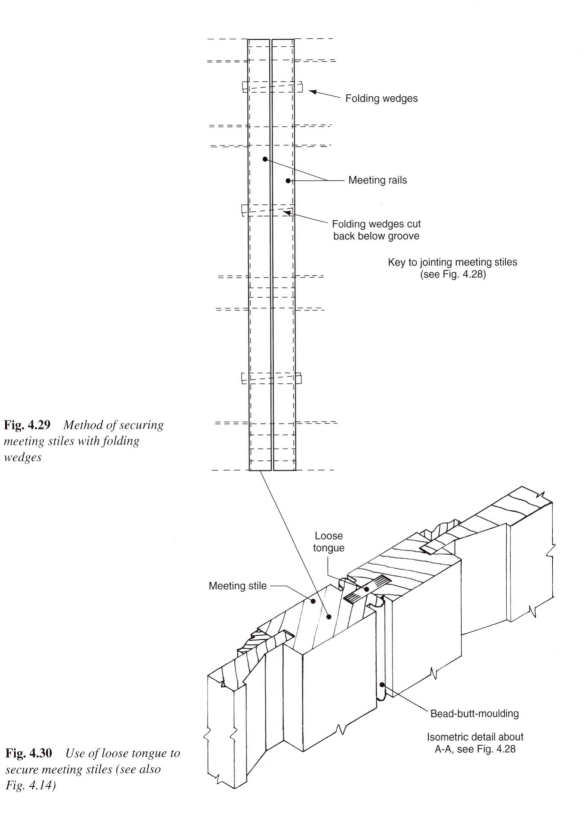

Fig. 4.29 *Method of securing meeting stiles with folding wedges*

Folding wedges

Meeting rails

Folding wedges cut back below groove

Key to jointing meeting stiles (see Fig. 4.28)

Loose tongue

Meeting stile

Bead-butt-moulding

Isometric detail about A-A, see Fig. 4.28

Fig. 4.30 *Use of loose tongue to secure meeting stiles (see also Fig. 4.14)*

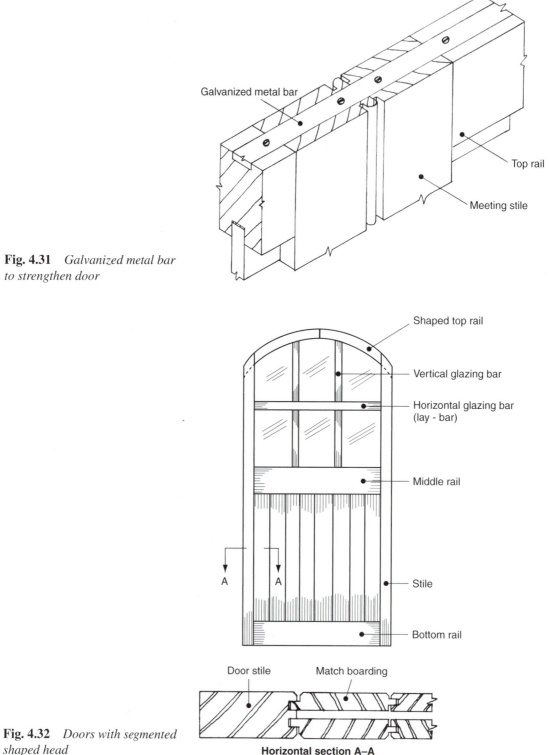

Fig. 4.31 *Galvanized metal bar to strengthen door*

Galvanized metal bar

Top rail

Meeting stile

Shaped top rail

Vertical glazing bar

Horizontal glazing bar (lay - bar)

Middle rail

Stile

Bottom rail

A   A

Door stile   Match boarding

Horizontal section A–A

Fig. 4.32 *Doors with segmented shaped head*

Next, the top rails, frieze rails, middle rails and bottom rails are inserted into the meeting stiles followed by all the panels. Finally, the outer stiles are secured and the whole door is cramped and checked for square before all the outer wedges are inserted. Dowels are also inserted in the joints for further strength.

After the door has been reduced in height, galvanized metals bars 25 × 6 mm are housed into the edges (which are stopped 12 mm from each end) at the top and bottom of the door and secured with non-corrosive screws (Fig. 4.31). This will help strengthern the two leafs

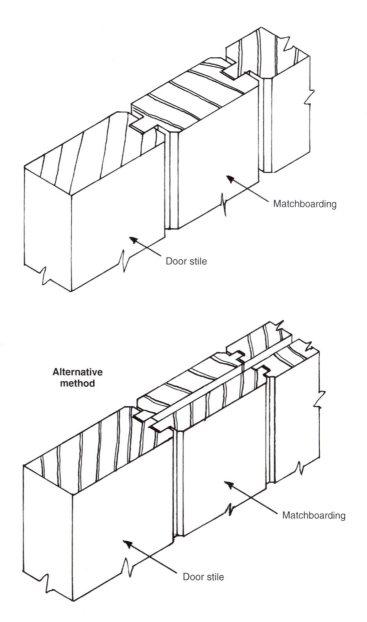

Fig. 4.33 *Matchboarding similar thickness to door*

Fig. 4.34 *Matchboarding both sides*

and prevent the door from sagging. One and a half pairs of 100 mm hinges are used to hang the door.

4.9 Doors with curved top rail

Figure 4.32 shows an example of a door with a segmented top rail suitable for internal use, where both sides of the door have been faced with tongue-and-groove boarding (matchboarding) below the middle rail.

Any joinery that involves circular work often requires the production of a setting-out rod to produce the curved work full size to enable templates and jigs to be made for use with woodworking machinery.

For doors of similar design used externally, the matchboarding would run past the bottom rail for weathering purposes similar for a framed-ledged-and-braced door (Fig. 4.8) with the matchboarding on the external face only. Volume 2, section 4.1 gives examples of the methods used to set out various geometrical shapes.

4.9.1 Construction

Again, mortise-and-tenon joints are used for the rails into the stiles with the matchboarding supported by a groove running around the inside edges of the door and below the middle rail similar to Fig. 4.8. Above the middle rail are vertical and horizontal glazing bars (horizontal glazing bars are sometimes referred as 'lay-bars'), which are stub mortised into the outer frame and cross-jointed, similar to those in Figs 3.22c to f. The matchboarding in this case is the same thickness as the door (Fig. 4.33) for decorative purposes and for easier construction.

Figure 4.34 shows an alternative method where two layers of matchboarding, approximately 18 mm thick, are used back to back.

Forming the curved top rail will require two or more pieces of timber (depending on width and shape of the door) that are shaped and jointed together on their ends with handrail bolts and dowels (see Volume 1, Fig. 12.11).

Jointing of the top rail to the stiles needs to be carried out taking into account the sufficient strength required at this point. One of two methods is recommended:

- a bridle joint with dowels (Fig. 4.35); or
- a 'hammer-headed' joint (Fig. 4.36) with hardwood wedges, which are inserted after assembly to pull up the shoulders.

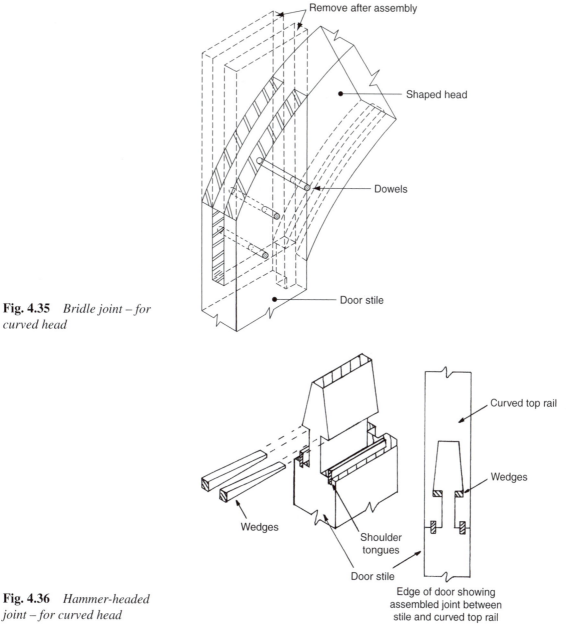

Fig. 4.35 *Bridle joint – for curved head*

Fig. 4.36 *Hammer-headed joint – for curved head*

Both methods will ensure sufficient strength but the latter may prove to be more expensive to construct.

4.10 Fire doors

Fire doors are required to close openings within walls and, to satisfy the Building Regulations, are required to offer fire resistance for a set minimum period of time.

If fire breaks out within a building, a fire door will have to perform two main functions:

(i) help contain the fire within a confined area, while allowing persons to safely vacate the building;
(ii) help prevent fire spreading and limit the movement of smoke to other areas.

Fire doors must therefore be self-closing at all times.

The design of these doors (and of the door frames which house them) is critical, in that they have to undergo stringent tests by an approved testing body in accordance with recommendations laid down within BS 8214: 1990 (code of practice for fire door assemblies with non-metallic leaves) and BS 5588–0: 1996 to ensure that their prescribed duration of fire resistance is met.

The current official term used by British Standards and the Building Regulations to describe these doors is 'fire-resisting doors' (BS 8214:1990) and is often preceded by figures indicating (in minutes) their integrity – the period of time during which a door resists the passage of flame or hot gases, or when flaming appears on the face of the door away from the fire.

4.10.1 Fire-resistant grading

Fire door manufacturers must now comply with the Building Regulations and relevant British Standards to identify fire-resisting doors by the following specification bodies:

- BWF (British Woodworking Federation);
- TRADA (Timber Research and Development Association).

4.10.1.1 British Woodworking Federation (BWF)

The fire-resistance of a fire door certified by the BWF is indicated by a label (Fig. 4.37). These labels are placed in specific locations, i.e. on the top of the door or on the underside of the frame head (Fig. 4.38a to d).

A foil-backed card ('gaptester' see Fig. 4.37) is also available to check the gap around the fire door as well as for checking for the appropriate labels on the top of the door.

4.10.1.2 TRADA (Timber Research and Development Association).

Fire-resisting doors certified by BM TRADA (Timber Research and Development Association) are coded by means of a small (9 mm

Note: *The letters FD (Fire-Door) are placed in front of the time period to identify it as a fire-resisting door and may also include the letter 'S', which indicates that a smoke seal is included*

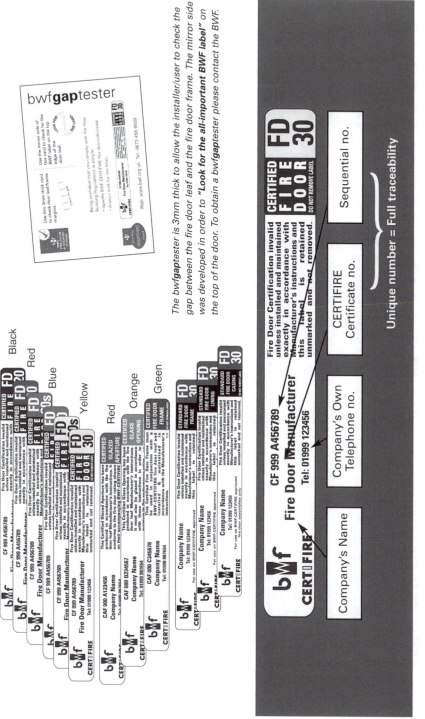

Fig. 4.37 *BWF fire-resistance labels*

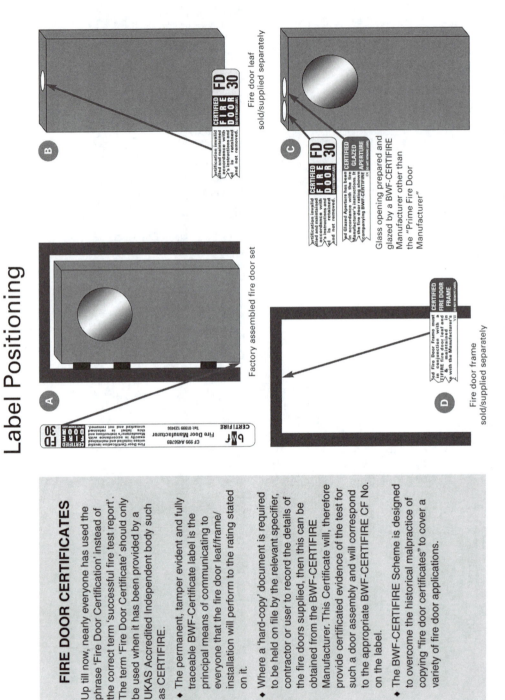

Fig. 4.38   *Position of BWF labels on doors and frames*

diameter) coloured cylindrical plug (Fig. 4.39) inserted into a 20 mm deep hole drilled into the edge of the hanging side of the door – positioned either 600 mm down from the top or 600 mm up from the bottom of the door.

Each plug (Fig. 4.39) has a coloured outer circle that identifies the fire-resistance of the door, which may be 30, 60, 90 or 120 minutes as shown in Table 4.6.

Generally as a guide, a 44 mm thick door has a fire resistant rating of 30 minutes and a 54 mm door 60 minutes; furthermore, the letters FD (fire door) are placed in front to identify it as a fire-resisting door and may also include the letter 'S', which indicates that it has a smoke-seal included (see Approved Document B, Fire Safety: 2000).

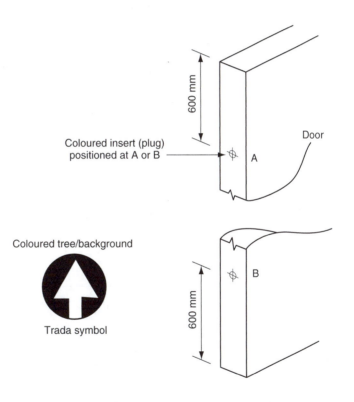

Fig. 4.39 *Position of BM TRADA fire-code plugs*

Table 4.6 Fire Resistance – Outer Colour Coding

Outer colour	Resistance against fire (minutes)
Yellow	30
Dark blue	60
Brown	90
Black	120

Independent fire door manufacturers include their membership number on the plug (Fig. 4.40) to prove that they have manufactured the door, using TRADA specifications.

The colour of the inner tree on the plug indicates whether an intumescent seal (see section 4.10.5) has been included with the door and/or frame to increase resistance against fire.

Other specifications may include glazing, whether it's a factory-hung door set or certified installation – see Table 4.7.

4.10.2 Construction

The construction of a fire door is critical to its performance. Therefore the design of the structure of a fire door should be left to manufacturers with experience in the manufacture of fire doors; or the services of an engineer with experience in fire safety engineering should be sought.

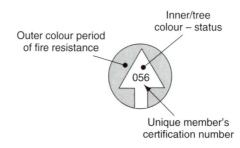

Fig. 4.40   *BM TRADA colour codes for plugs*

Note: *Intumescent seals with smoke seal must be fitted in specified areas. See Building Regulations, Approved Document B, Fire Safety; 2000 – Table B1 – Provisions for fire doors.*

Table 4.7   Inner/tree colour and specifications – when plug is fixed to door edge

Fire door type	Inner/tree	Additional protection	Inner/tree colour
Approved FD30 & FD60 only. Intumescent seals to be fitted.	Red	Approved factory fitted glazing	Mustard
Approved FD90 & FD120. Intumescent seals in door – factory fitted.	Green	Certified factory hung door set.	Silver
		Certified installed door set.	Lime
When plug is fixed to door frame			
Approved frame to match door. All intumescent seals to door and frame fitted.	Green	As above	

Specialist manufacturers nowadays tend to use manufactured boards that offer a high resistance to fire. These tend to be a lot heavier than conventional doors, thus requiring good sturdy hinges for support – often three to a door with the central hinge nearer the top (see section 4.14.4).

Figure 4.41 shows a typical part horizontal section through a modern manufactured fire-resisting door rated as an FD30. An intumescent seal (see section 4.10.5) is now included in all fire-door sets and may be placed on the edge of the door, or around the inside edge of the frame. As stated previously, all such designs must be tested before any rating is approved.

> Note: In order to be impartial these definitions should not refer to the BWF or to certificated manufacturers

4.10.3 Definitions relating to fire resistance (taken from BWF)

- *Fire door* – a door-leaf design which has been tested to resist fire and the main components of a fire door assembly;
- *Fire door frame* – a door frame made for a fire door which must be compatible with the fire door;
- *Glass opening* – a special opening in a fire door produced in preparation for glazing;

> Note: If the glazed opening is not carried out in accordance with the door manufacturer's instructions, the fire door's certification is invalid

- *Glazed aperture* – an opening glazed in accordance with the door manufacturer's instructions using the correct fire-resisting glass, beads and intumescent materials;

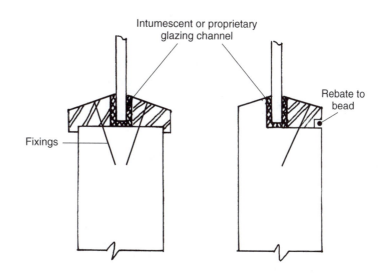

Fig. 4.41 *Possible arrangements for beading glass into fire-resisting doors*

- *Fire door-set* – a fire door-leaf which is supplied pre-hung, in (or along with) it's compatible fire door frame, hinges, glazing, intumescent fire (and smoke) seals and door furniture from one source;
- *Fire door assembly* – all the correct compatible components, although supplied by different manufacturers;
- *Fire door installation* – a fire door set or fire door assembly that has been installed satisfactorily to comply with the Building Regulations and relevant British Standards.

4.10.4 Glazed apertures

These are permitted within a fire-resisting door or screen, provided that the glass is of a type proved by fire test, and is limited to an area and aspect ratio as defined in the fire door test report.

The method of fixing the glass is very important and should be in accordance with BS 8213–4:1990 and the door and glazing manufacturer's instructions if failure around the edges is to be avoided. Figure 4.41 shows examples of how beads can be fixed.

4.10.5 Intumescent seals

An intumescent material is one that expands when subjected to high temperature. Such materials are available in paint, paste, or strip form.

If intumescent material in strip form is fixed into the top and side edges of a door, or within the rebate of its frame, then if that door is subjected to temperatures of between 140 and 300° (depending on the product type), the material will expand and seal the gap, which should not exceed 3 mm. At these temperatures the intumescent material will foam and expand to between four and five times its original volume – it will not burn. Figures 4.42a, b and c show various methods of application (there are others) in the door jamb and the edge of the door.

The result of the material being activated is shown in Fig. 4.42b and c.

Note: *The use of a 25 mm deep doorstop or rebate is no longer necessary or desirable. The inclusion of an intumescent seal reduces the use of a doorstop to its ability*

These intumescent seals are an integral part of the fire-resistance rating given to the door-leaf in conjunction with its approved frame.

Figure 4.42d shows where, in the case of an FD30 fire door, the intumescent seal may be interrupted by the hinge. However, with FD60 fire doors, two intumescent seals are required.

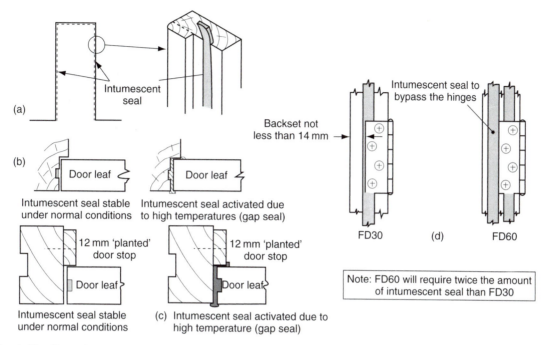

Fig. 4.42 *Door-frame-to-door treatment*

4.11 Door frames

Door frames require a rebate to act as a door check, otherwise the door would swing through the opening, straining the hinge and/or splitting the jamb. Methods of forming a rebate can be seen in Fig. 4.43. Door frames should be strong enough to carry a door without receiving support from a wall and should usually arrive on site ready assembled – with the exception of large openings (Fig. 4.44b).

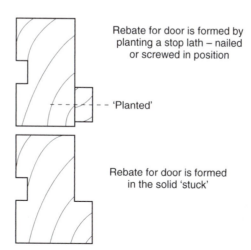

Fig. 4.43 *Forming rebate for door*

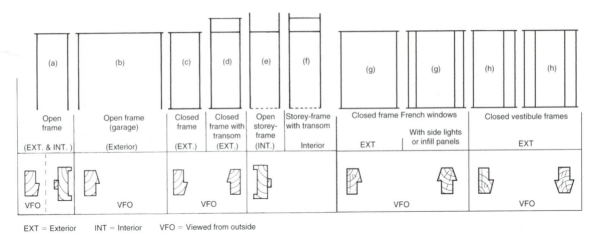

EXT = Exterior   INT = Interior   VFO = Viewed from outside

Fig. 4.44 *Door frames*

4.11.1 Open door frames (Figs 4.44a and b)

These are mainly used with exterior doors. Construction and erection details are shown in Fig. 4.45.

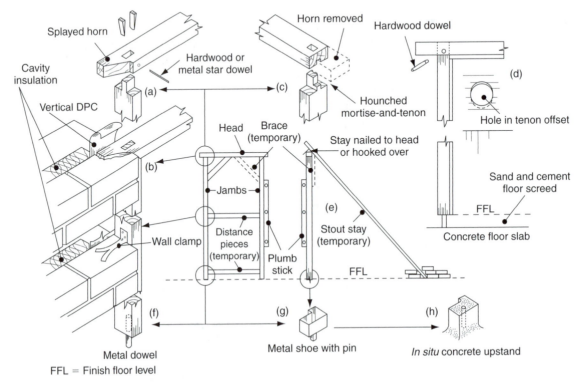

FFL = Finish floor level

Fig. 4.45 *Open door frames assembly and fixing*

4.11.2 Closed door frames (Figs 4.44c, d, g, h)

These are exterior door frames which may include in their design a sill (threshold) or 'fanlight' (glazed opening above the door) as shown in Figs 4.44c and 4.44d. Methods of providing these openings are shown in Fig. 4.47d.

In the case of French windows or Vestibule frames, Figs 4.44g and 4.44h, these may include side lights (glazed areas) to one or both side of the door opening.

Figure 4.46 shows how various frame members are arranged to suit either an inward- or an outward-opening door. If the main frame-work is constructed of softwood, then the sill should be made from a durable hardwood (see Volume 1, Tables 1.14 and 1.15). Corner joints may be mortised-and-tenoned or combed. Joints between the transom rail and jambs should be through mortised-and-tenoned. Adhesives should be at least BR types (see Volume 1, Table 11.1).

Frames are erected and fixed to masonry in much the same way as described for casement windows (see Chapter 3, section 3.6.1), i.e. building-in horns and the use of wall cramps. Fixing to pre-formed openings is also possible, as with door sets (described and illustrated in section 4.15).

High-performance features such as weather-seals within door frame rebates are also sound practice (Fig. 4.46a).

4.11.3 Door frame assembly

Figure 4.45 deals with the making, assembly, and fixing of a suitable open door frame. The head and jambs are joined together by using a mortise-and tenon joint (Figs 4.45a and b), which should be coated with paint or a suitable resin adhesive, assembled, cramped, wedged, and dowelled. If cramping the joint is not practicable, the joint could be draw-bored (Fig. 4.45d) – when a hardwood dowel is driven through the off-centre holes the shoulders of the joint will be pulled up tight. Because the frame has only three sides, a temporary tie (distance piece) fixed across the bottom of the jambs and a brace at each corner will be necessary if it is to retain its shape.

If the frame is to be built into the structure, temporary propping will be needed (Fig. 4.45e). When the frame has been accurately positioned – both level and plumb – the permanent securing process can begin. Firstly, provision is made at the foot of each jamb for good anchorage to the step, either by using a metal dowel (Fig. 4.45f), shoe (Fig. 4.45g) or a concrete stool (Fig. 4.45h). As the walls are built up on either side, wall clamps are fixed at approximately 500 mm intervals with screws to the back of the jambs (Fig. 4.45), then walled in. On reaching the head (Fig. 4.45a), it will be seen that the *horns* have been cut back on the splay. This allows the face

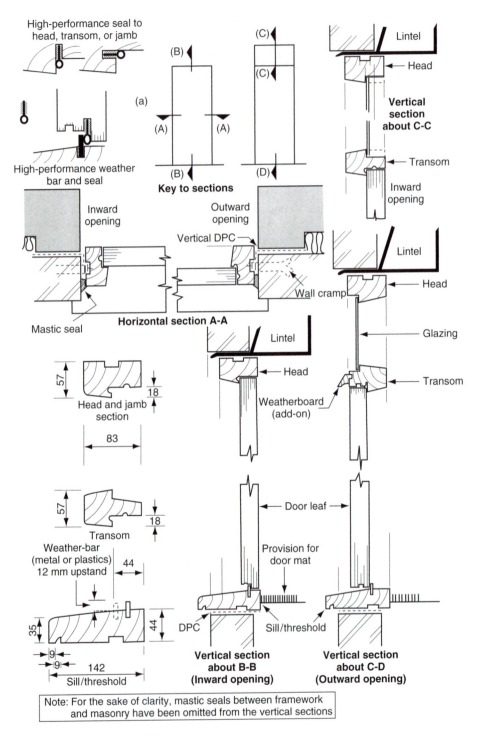

Fig. 4.46 *Sectional details through door frames*

walling to lap and totally enclose them, to make ready for the lintel above.

If, however, the frame is to be fixed into an existing opening, other fixing devices will have to be employed, for example wood (not recommended) or plastics plugs, wedges, etc. (see Volume 1, section 12). Either way, it follows that the horns will not be needed as a fixing aid and they will therefore have to be sawn off. For greater stability, a haunched mortise-and-tenon joint is then used (Fig. 4.45c).

4.11.4 Storey frames (Figs 4.44d, e and f)

As the name implies, with a storey frame, the jambs are at least as long as the room is high. The frame may incorporate a fanlight, which can be a very useful feature when borrowed light, ventilation, or both are required from an adjacent room or space. As shown in Fig. 4.44, the framework is made up of a grooved section. This groove provides lateral (sideways) support both to the frame and to the partition/wall that is set into it. As the storey frame is erected before the partition, it also provides a useful vertical guide for the blockwork, etc. which follows.

Figure 4.47a shows how frames which incorporate a fanlight are jointed. Those without are shown in Fig. 4.47b, where the space between the frame head and the ceiling has provision made so that it can be clad with plasterboard – possibly with an infill of insulation (sound and/or thermal) material, depending on the partition location.

Figure 4.47c shows how the jambs may be fixed to the ceiling joists. It should be noticed that whichever way the joists run provision can still be made for fixing, either by using packings or by inserting bridging pieces (noggings) between the joists.

Variations on how the fanlight area can be treated are shown in Fig. 4.47d.

4.12 Door linings (often referred to as door 'casings')

Linings are used internally on inner single-leaf walls to mask the reveal and soffit of a doorway. They must at the same time provide sufficient support by which an internal door can be hung. They may arrive on site requiring assembly before fixing, but can be already pre-assembled beforehand in the workshop.

Section sizes vary but are generally smaller than those used for door frames and because of this they are often made using housing joints (Fig. 4.48) – mortise-and-tenon joints being unsuitable. When fixed, the outer edges of the lining are in line with the finish of the wall, which may be plaster, or any other type of suitable wall covering.

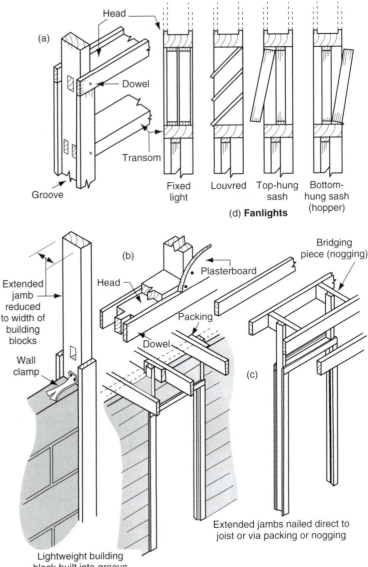

Fig. 4.47 *Storey frames – within partition walls*

Walls of brick or block walls are often left as a natural feature with no finish.

Linings are constructed to suit the thickness of the internal wall and are classified as either narrow or wide linings.

4.12.1 Narrow door linings

Figure 4.48 shows how both solid (rebated) and plain (planted-rebate) linings are sited in relation to the wall, and how the joints between the jambs and head are formed and assembled.

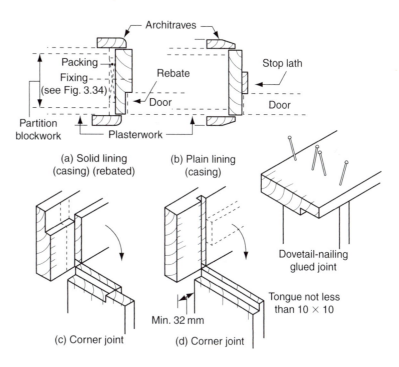

Fig. 4.48   *Narrow door linings (casings)*

Figure 4.49 shows the procedure for fixing an assembled door lining into a pre-formed opening (see section 4.13.1, Framing openings). Fixings may be made by:

(a)   proprietary fixings (see Fig. 3.34);
(b)   driving wood plugs into brickwork seams (not recommended, see Volume 1, section 12.1);
(c)   using cut clasp nails driven into lightweight building blockwork;
(d)   plastics plugs and screws – this method is not suitable for frames which house fire doors, because of the low melting point of plastics.

Note: *A lining must be in contact with its wall at all fixing points (not to exceed 600 mm centres), either directly or via a packing piece*

4.12.2 Wide door linings

Because of the impracticality of using wide solid timber, framed panels (Fig. 4.50a) or a skeleton framework clad with board (Fig. 4.50b) can be used to construct wide door linings. In both cases they will be fixed to a framework of timber grounds (Fig. 4.50c) pre-made in the workshop to suit the reveal opening. Any framing joint may be used – Fig. 4.50d shows some examples.

In high-class work the lining will often include a plinth block to provide a neat finish where the skirting meets the base of the architrave (Fig. 4.50).

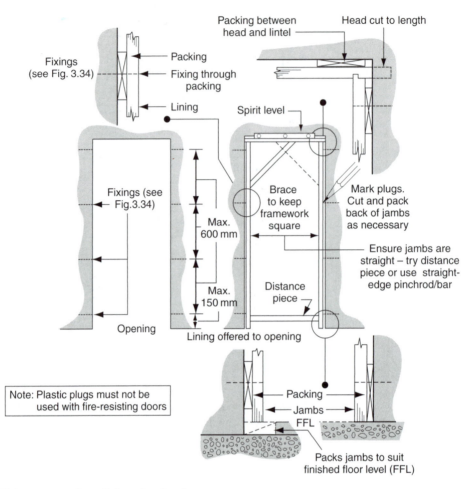

Fig. 4.49 *Fixing narrow door linings (casings)*

Intro

Table 4.8 shows a door schedule, which should be read with the sample floor plans to a domestic house shown in Fig. 4.51.

4.13 Door schedule (see Table 4.8 and Fig. 4.51)

Before any door is fitted and hung, the door schedule needs to be consulted. This is a visual means of identifying the fitting requirements of each individual door and will cover such areas as:

   (i)   type of door;
  (ii)   location;
 (iii)   sizes;
 (iv)   door handing (see Fig. 4.53);
  (v)   hardware required;
 (vi)   any glazed openings;
(vii)   finish to door.

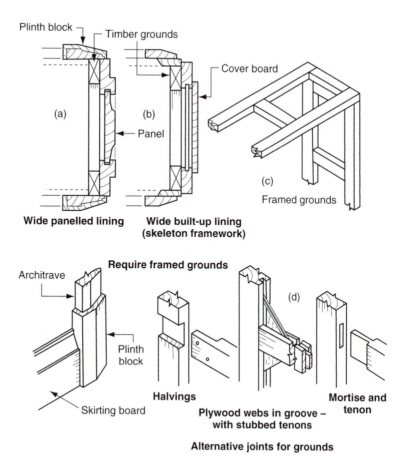

Fig. 4.50 *Wide door linings*

Also it is used in the planning stages of the construction to identify the different components required. This should ensure that all items are ordered and available when the door is fitted.

4.14 Door hanging (door types)

Doors are used to seal off rooms or areas for security (especially in external walls), as well as for privacy (particularly in bedrooms and bathrooms), even if the door is slightly ajar (Fig. 4.52). The door hanging side is determined by the architect's floor plans where consideration is given to the required maximum floor space and the area taken up by the door when in the open position, i.e. space for kitchen fitments, lounge furniture, etc. Furthermore, the location of the light switch needs to be easily assessable when entering the room. Where doors have been altered at a latter stage to hang on the opposite side (for convenience), the light switch is no longer easily assessable and will need to be relocated.

The door frame or lining, as previously described, must be securely fixed to the structure to take the weight of the door, especially when

Table 4.8 Door hardware (ironmongery) schedule – refer to Fig. 4.51 (Location plan of building)

Description	Location (Fig. 4.51)										Total
	D01	D02	D03	D04	D05	D06	D07	D08	D09	D010	
Door											
Hardwood Ex – Circular headed entrance door 2032 × 838 × 44 – Type EC227	1										1
Softwood Ex – French window lights 1981 × 584 × 44 – Type ES479						2					2
Softwood Ex Stable door 1981 × 762 × 44 – Type IG947					1						1
Softwood Int. Flush panelled effect door 1981 × 762 × 35 – Type IP364		1	1	1			1	1	1	1	7
Hardware											
100 mm Brass butts hinges – knuckle ball bearings Cat. No. BB/BB 02 (pairs)	1										1 pr
100 mm Brass Butts – washed hinges Cat. No. BB/BB 06 (pairs)					1	3					4 prs
75 mm Steel butts hinges. Cat. No. SL/65 (pairs)		1	1	1			1	1	1	1	7 prs
65 mm 5 Lever Rebated upright mortise sash lock (BS EN 1300:2004) Cat. No. SL/66						1					1
65 mm 5 Lever upright mortise lock (BS EN 1300:2004) Cat. No. SL/62					1						1
65 mm 5 Lever upright mortise Dead lock (BS EN 1300:2004) Cat. No. SL/60						1					1
75 mm 5 Lever upright mortise Sash lock (BS EN 1300:2004) Cat. No. SL/67	1										1
Night latch with brass cylinder Cat. No. NLB64	1										1
75 mm Tubular latch Brass plated Cat. No. TLB88		1	1	1			1	1	1	1	7
Brass plated door chain. Cat. No. BP381	1										1
180° Security door viewer. Cat. No. SDV189	1										1
Brass Letter plate 255 × 75 mm. Cat. No. BLP78	1										1
Brass Lock Lever door furniture set Cat. No. BF/678	1				1						2
Brass Lever latch door furniture set Cat. No. BF/343		1	1	1			1	1	1	1	7
Brass Indicator bolt Cat. No. IL/B56	1										1
Security hinge bolts Cat. No. SHB/731 (pairs)	1				2	2					5 prs
Brass Security barrel bolt Cat. No. HB/56					1	2					3
Surface finish											
Painted – Make: Well-known Ltd. Colour Code – Type 56DC		X	X	X	X		X	X	X	X	8
Self Finish – (Translucent) Make: Well-known Ltd. Code – Type T45	X										1
Stained – Make: Well-known Ltd. Code – Type ST89						X					1

Range

IP364   ES479   IG947   EC227

Note: *As shown in this example, references have been made to a catalogue (Cat.) number – in this case a supplier/manufacturer should be substituted*

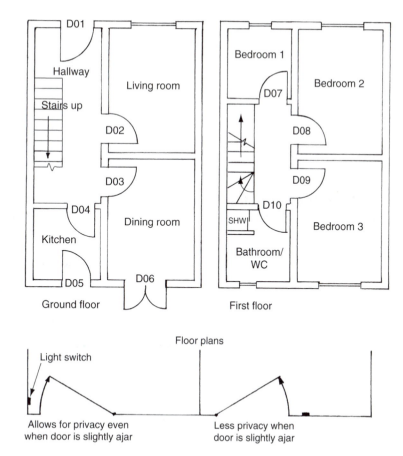

Fig. 4.51 *Sample floor plans to a domestic house*

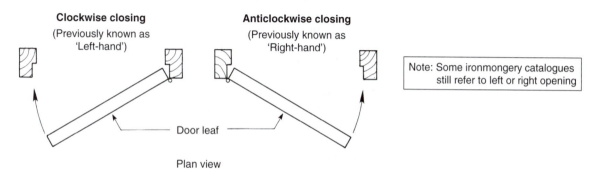

Fig. 4.52 *Positioning the hanging side of the door*

the door is in the open position – this puts extra strain not only on the hinges, but also on the frame fixings.

The term used to describe the side and edge of a door frame from which a door-leaf is hung is 'handing'. Two methods of handing are shown in Fig. 4.53.

Door clearance is shown in Table 4.9.

Fig. 4.53 *Handing for door hanging and hardware*

Table 4.9   Door-leaf edge clearance

Frame/lining member	Door-leaf edge clearance (mm)	
	Internal doors	External doors
Jamb leg	2	2.5
Head	2	2.5
Transom	2	2.5
Threshold	3	3.5

4.14.1 Ledged-braced-and-battened doors

When hanging this type of door the width of the rebate on the frame will be determined by whether the door is to swing outwards or inwards, i.e. whether the hinges (tee-hinges) are to be fixed onto the ledges or directly onto the matchboarding. Tee hinges (Fig. 4.69) are used as butt hinges would be difficult to fit to the edge of the door (Fig. 4.54).

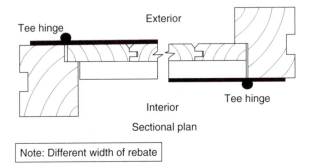

Fig. 4.54   *Positioning the door*

Once the method of hanging and 'handing' has been established the next operation is to fit the door. The following is guidance on how this can be achieved.

1. Check that the bottom of the door is parallel with the step or threshold. Fit it if necessary.
2. Fit the door's hanging edge into its rebate, and ensure it is straight with no gaps.
3. a. With an assistant, position the door's hanging edge just inside its rebate. Using half the width of a pencil, scribe the closing edge. This will provide door clearance (Fig. 4.55). Remove waste wood and bevel back to produce a 3 mm lead-in – known as a 'leading edge'.
   b. Alternately, measure the distance between rebates (allowing for clearance), transfer it to the door's face, and remove waste wood as in 3a.
4. Place the door into the rebates, remembering to allow for floor clearance. Mark its height from inside the top rebate and remove waste wood.

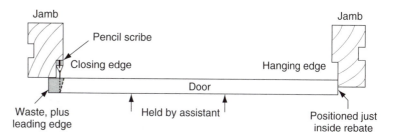

Fig. 4.55   *Fitting the door to the closing edge*

5. Lay the door flat over two saw stools and screw tee hinges (Fig. 4.69) to the battens or ledges, using only two screws per hinge.
6. Position the door into its frame. Use thin wedges as packings and adjust to give clearance all round (2–3 mm at the sides and top, 4–6 mm to the floor). Fix the top hinge to the jamb, using only one screw, then fix the bottom hinge likewise.
7. Remove the packings. Check the clearance, noting any adjustments needed, and remedy if necessary.
8. Repack the door. Unscrew the door from its frame (bottom hinge first), then remove both hinges from the door.
9. Paint the backs of the hinges and refix them to the door, using *all* the screws this time.
10. Rehang the door and fix all the remaining screws.

Note: *When using pre-painted hinges, stages 9–10 will not apply*

4.14.2 Fitting hinges to standard doors

Framed doors, i.e. framed-ledged-and-braced, panel, flush, etc., allow the butts to be fitted to the edge of the door rather than the face. Hanging the door is similar, but fitting the butts requires them to be 'housed' into the edge of the door and frame. This may be carried out by the traditional method, or by more modern methods with portable power tools and jig (section 4.14.3).

The stages which follow should be read in conjunction with Fig. 4.56 (*Hanging a door with standard (cranked) butt hinges*).

1. Position the hinges on the door edge, 150 down and 225 up, or see Fig. 4.57. Mark their position.
2. With the door positioned within the frame (check the top clearance), use a straight edge to transfer the hinge positions on to the frame or lining.
3. Set a marking gauge or combination square to the width of the hinge leaf.
4. a. Gauge or rule the hinge-leaf width along the door edge and the frame or lining. Where rebates have been formed from the solid, stage 4b will apply.
   b. With frames and linings with stuck rebates, provision must be made for rebate back clearance. A back template made from

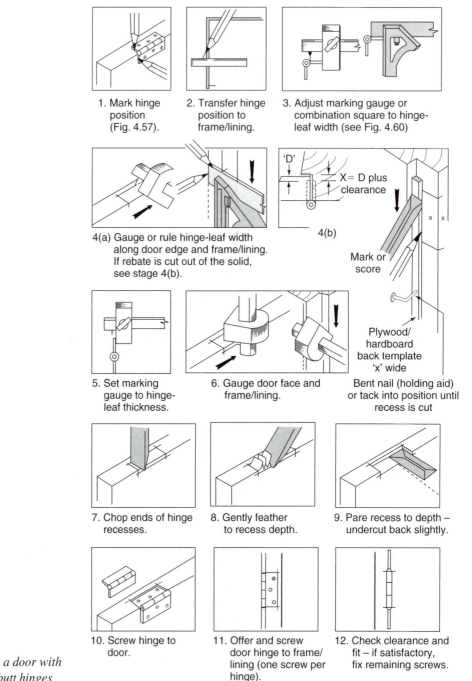

Fig. 4.56 *Hanging a door with standard (cranked) butt hinges*

1. Mark hinge position (Fig. 4.57).

2. Transfer hinge position to frame/lining.

3. Adjust marking gauge or combination square to hinge-leaf width (see Fig. 4.60)

4(a) Gauge or rule hinge-leaf width along door edge and frame/lining. If rebate is cut out of the solid, see stage 4(b).

4(b) 'D' X = D plus clearance Mark or score Plywood/ hardboard back template 'x' wide Bent nail (holding aid) or tack into position until recess is cut

5. Set marking gauge to hinge-leaf thickness.

6. Gauge door face and frame/lining.

7. Chop ends of hinge recesses.

8. Gently feather to recess depth.

9. Pare recess to depth – undercut back slightly.

10. Screw hinge to door.

11. Offer and screw door hinge to frame/ lining (one screw per hinge).

12. Check clearance and fit – if satisfactory, fix remaining screws.

hardboard or plywood cut to width 'x' (door thickness minus gauged hinge-leaf width plus back clearance of 1 to 1.5 mm) will enable a line to be marked with a pencil or scored with a scriber or chisel.

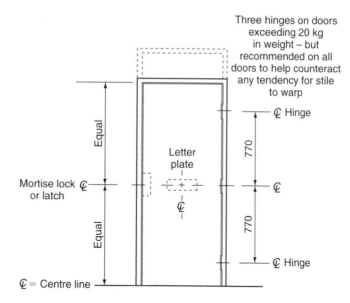

Three hinges on doors
exceeding 20 kg
in weight – but
recommended on all
doors to help counteract
any tendency for stile
to warp

℄ Hinge

770

Letter
plate

Mortise lock ℄
or latch

℄

770

Equal

Equal

℄ Hinge

Fig. 4.57 *Positioning hardware on doors, frames, linings and door sets*

℄ = Centre line

5. Set the marking gauge to the hinge-leaf thickness.
6. Gauge the door face and the frame or lining edge over the length of each hinge-leaf position.
7. Holding the chisel vertical, accurately chop the ends of each hinge-leaf recess to the gauged line.
8. Gently feather to the recess depth with the chisel sloping, its face uppermost.
9. Pare and clean each hinge recess. Undercut very slightly towards the back.
10. Fully screw each hinge to the door, after first using a bradawl to form a pilot hole. If slot-headed screws are used, the slots should be in line with the door height – in this way, paint or polish build-up is avoided and appearance is enhanced.
11. Offer the hinged door to the frame or lining recess. Pack-up as necessary and, using one screw per hinge, secure in position.
12. Check the door leaf for clearance and fit. If satisfactory, fix all remaining screws.

4.14.2.1 Using a 'Stanley' butt gauge

When hanging a door leaf into a *non-rebated lining,* this provides an alternative to using a marking gauge at stages 3 and 5 above. (Rebates are formed with planted stop lath after the door leaf has been hung.)

1. Using the gauge as a square, transfer the face positions of the hinge leaf to the door-leaf edge and to the edge and inside face of the lining (not shown) (Fig. 4.58a).

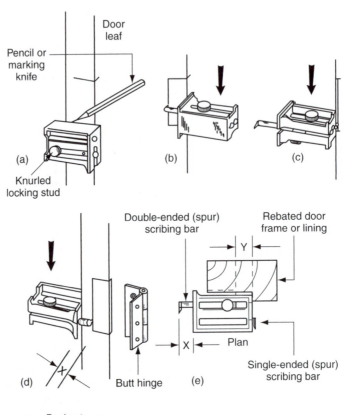

Fig. 4.58 *Application of a*
'Stanley' butt gauge

x = Backset
y = Backset (x) plus door clearance (set automatically)

2. Set the leading spur of the double-ended scribing bar to the width of the hinge leaf. Scribe the edge of the door from the door-leaf face and the face of the door lining from its edge (Fig. 4.58b).
3. Set the single-ended scribing bar to the hinge-leaf thickness (cranked butt hinge), then scribe the face of the door leaf and the edge of the door lining (Fig. 4.58c).

With a rebated door frame or lining, stages 1 and 3 remain the same but, instead of setting the spur to the width of the hinge leaf as in stage 2, make the following modifications:

(a) Remove the double-ended scribing bar from its stock and reposition it upside down, so that both spurs are exposed. (Part of the scribing bar will be visible at one end, while only the point of the spur can be seen at the other end.)
(b) Set the spur of the protruding scribing bar to the distance (x) remaining across the door-leaf edge, as if the hinge were in position

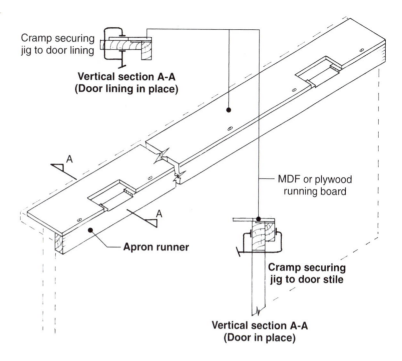

Cramp securing
jig to door lining

**Vertical section A-A
(Door lining in place)**

MDF or plywood
running board

A

A

Apron runner

**Cramp securing
jig to door stile**

**Vertical section A-A
(Door in place)**

Fig. 4.59 *Isometric and sectional details of a purpose made door hinge jig*

as shown in Fig. 4.58d. In this way, the spur of the rear scriber is automatically set to include door clearance between the door face and rebate check when scribing from within the door frame or lining rebate, as shown in Fig. 4.58e.

4.14.3 Door hinge jigs

Jigs of this type are used in conjunction with a portable power router, to recess the hinge leaf into a door edge and its casing or frame. They may be purpose made or a proprietary product.

A purpose-made jig suitable for an internal door and its lining (with planted rebate) is shown in Fig. 4.59. It consists of an MDF or plywood running-board template with cut-outs to suit the surface dimension of the hinge leaf plus, as shown in Fig. 4.60, an offset to allow for the thickness of a guide-bush and cutter clearance (see also Volume 1, Fig. 6.44). The jig is held secure to the door with cramps via a timber apron runner fixed to the underside of the running-board at a distance from the face to suit the hinge type. The door likewise must be held vertically secure on its edge, whereas the hinged side of the lining will require securing to the jig horizontally. Each routed hinge-leaf recess will result in two rounded corners; these can either be cut out square with a chisel, or by using a proprietary corner chisel, which after positioning is struck with a hammer.

Note: *This type of jig/template is generally only suitable where several doors of the same type are to be hung*

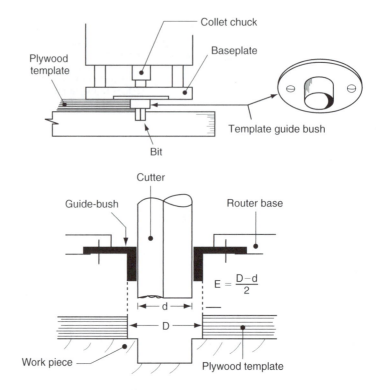

Fig. 4.60 *Making provision for the guide-bush and cutter when out the template*

Note: Correct PPE must be used as well as a suitable extraction system

Fig. 4.61 *Using the 'trend' hinge template*

The 'trend' hinge jig as shown in Fig. 4.61 is a portable metal template for recessing the hinge leaf into the door edge and inside face of its reveal (casing or frame) – it can be used for both standard and fire door requirements.

| (a) Setting the length | (b) Setting the width | (c) Setting the depth of leaf thickness |

Fig. 4.62 *Making preparations for using a 'trend' door hinge jig on a door stile*

Figure 4.62 shows:

(a) the hinge-leaf aperture being set to length;
(b) the hinge-leaf aperture being set to width;
(c) how provision is made for depth of cut to accommodate the thickness of leaf.

4.14.4 Fire doors (see section 4.10)

The construction, hanging and fitting of fire doors is strictly controlled by current building regulations and fire doors are only constructed by certificated manufacturers.

Once the door has been fitted the required hardware is as follows:

Hinges: these should have narrow leaves and be made of non-combustible material (having a melting point higher than 800°C). The positioning and number of hinges will depend on the weight of the door and the registered fire door manufacturer's specifications – usually three is specified. Figure 4.63 gives an example.

Closer: all fire doors will require an approved self-closing device (Figs 4.87 and 4.88). Rising butts (Fig. 4.76) are allowed for this purpose only in certain situations.

> Note: *Hardware should not interfere with intumescent strips but, where this is unavoidable, such areas may be coated with an intumescent paste*

Mortise locks/latches: like the hinges, these should be as slender and/or low-mass as practicable, thus reducing the amount of metal which will conduct heat in the event of a fire.

4.15 Door sets

A door set can be regarded as an external or internal door which has been matched to a frame or lining, assembled, and pre-hung under factory conditions with all the necessary hardware – locks, latches, etc. It will be delivered to site either fully assembled or packaged in such a way that it can be assembled as and when required (Fig. 4.64).

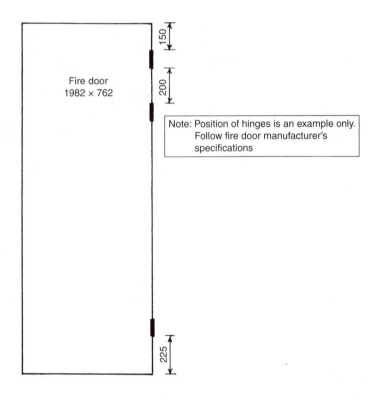

Fig. 4.63   *Examples of positioning hinges on fire door*

Because the door set may have been surface-finished in the factory (painted or polished), the use of detachable hinges such as the 'snappinge' – snap-in door hinge (Fig. 4.64). This snap-in mechanism permits easy assembly and dismantling of door sets. The hinge knuckle portion is housed in the door jamb, and the pocket into which the tongue fits is housed into the door leaf as shown in Fig. 4.64a. Because the pocket does not protrude beyond the face of the door, doors can be stacked flat one upon the other without interference of the hinge knuckles. Figure 4.64b shows how the hinge tongue is released from its pocket when the door leaf is removed from its frame enables the door leaf to be delivered at a later stage, thereby reducing the risk of surface damage (see also Figs 4.71, 4.72 and 4.76).

4.15.1 Framing openings

Templates (profiles) 10 mm wider and taller than the door set are built into the masonry to leave a slightly oversize opening on their removal. Templates should be detachable, to facilitate removal.

Figure 4.65 shows two types of template. The system shown on the left-hand side consists of a detachable framework lined with 12 to

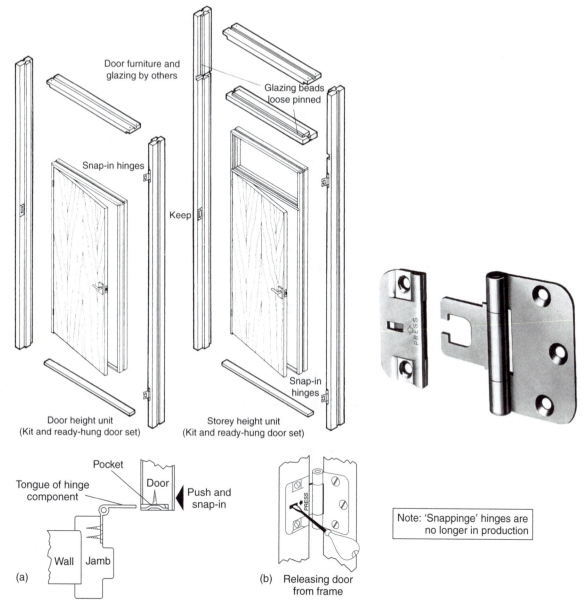

Fig. 4.64   *Door set system*

18 mm plywood of a width equal to the thickness of the wall plus any plasterwork. The plywood liner is fixed to the blockwork, etc., as the wall is being built, and acts as a profile for the plasterer. After the inner framework is removed (when the masonry is complete), the plywood is retained as a fixing for the door set.

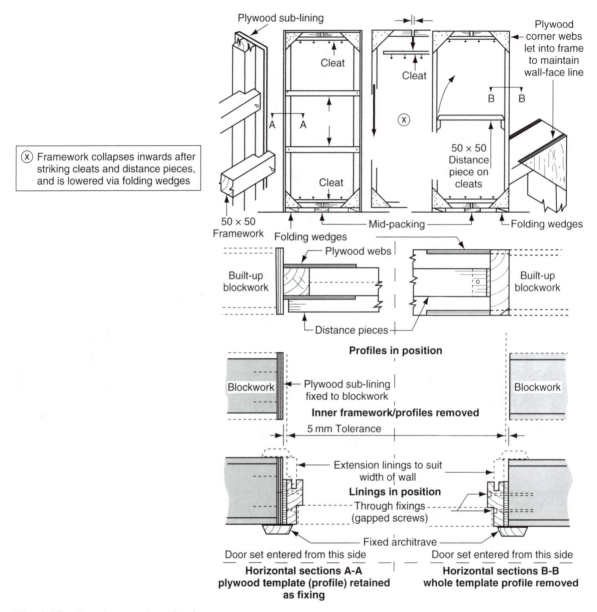

Fig. 4.65   *Forming openings for door sets*

The system shown on the right-hand side of Fig. 4.65 involves removing the whole (detachable) template to leave an opening of fair-faced blockwork, etc. In this case, the door frame or lining will need fixing to the wall via pallets or plugs and screws. (Plastics plugs must not be used as a fixing for a fire-door set.)

In the example shown in Fig. 4.65, the door set linings are made out to suit various wall widths by using add-on extension pieces.

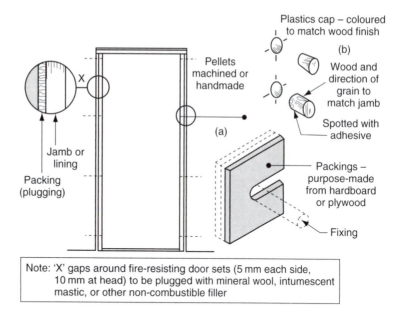

Note: 'X' gaps around fire-resisting door sets (5 mm each side, 10 mm at head) to be plugged with mineral wool, intumescent mastic, or other non-combustible filler

Fig. 4.66 *Fixing points for frames, linings and door sets*

4.15.2 Fixing a door set into masonry walls

(i) If necessary, assemble the door set framework/lining (Fig. 4.64), including one set of architraves.
(ii) Drill the door frame/lining to receive fixing screws – see Fig. 4.66.
(iii) Offer the frame/lining to the opening – the fixed architrave acts as a check (restraint).
(iv) Level the head or transom rail – adjust the foot or framework as necessary.
(v) Check for straightness and plumb.
(vi) Using the holes in the jambs, drill into the masonry. Insert plugs and screws – see Chapter 3, Figs 3.33 and 3.34.
(vii) Position packings (Fig. 4.66a) behind all fixing points between the back of the jambs and the wall. Tighten the screws. Check with a distance piece or the door leaf (on drop-on or push-in hinges) for correct clearances.
(viii) Remove the door leaf (if supplied). Plug all fixing points with wood pellets or plastics caps, as shown in Fig. 4.66b.
(ix) Fix loose architraves on the opposite side (Fig. 4.67) and the fanlight if fitted (see also section 8.2).

With the exception of stage (vi), the above fixing procedures will also apply to timber-studded partitions.

Note: *Fire-door sets should have any gaps between the jambs and the wall plugged with a non-combustible filler as shown in Fig. 4.66*

4.16 Domestic sliding interior doors

Sliding a door across an opening provides a useful alternative to hinging where floor space is restricted. There is, however, the disadvantage that unless the door slides into a wall, surface wall space is lost.

There are several proprietary systems on the market. In the main, these consist of an overhead track from which the door is suspended and slide via a series of wheels made from plastics. The whole of the top sliding gear is covered by a pelmet, which should be made detachable

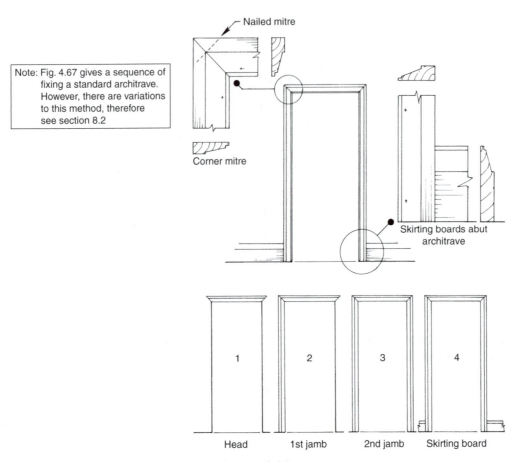

Note: Fig. 4.67 gives a sequence of fixing a standard architrave. However, there are variations to this method, therefore see section 8.2

Nailed mitre

Corner mitre

Skirting boards abut architrave

1   Head

2   1st jamb

3   2nd jamb

4   Skirting board

Fig. 4.67   *Sequence of fixing architraves (see also Fig. 8.14)*

to enable maintenance to the gear to be carried out without disturbing any decorations.

The foot of the door may be grooved to receive a metal channel which, as the door is slid open or shut, runs over a plastics guide-pin secured to the floor.

Figure 4.68 shows a typical arrangement for sliding-door gear.

In large door openings it may be possible to have a pair of sliding doors running on the same track and abutting when closed (see Chapter 10, Joinery fitments).

Alternatively, instead of the doors running across the face of the wall, they may disappear into specially constructed stud partitioning either side.

4.17 Storage and protection

Doors, their frames, and door sets should not be delivered to site until they are required. If, on delivery, it is found that they have not been

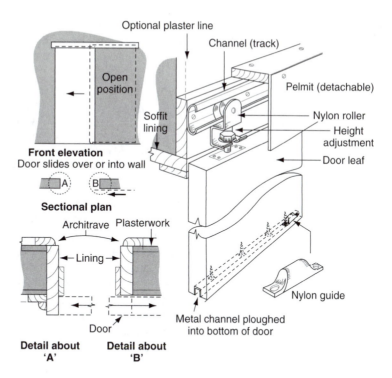

Fig. 4.68   *Domestic sliding-door gear*

factory-sealed or primed, both faces and all four edges should be immediately treated with the appropriate sealer or primer.

All doors awaiting use should be close 'stacked' (not less than three bearers placed across the length of each door) and stacked flat (out of twist) off the floor and under cover of a well-ventilated building.

Door stacks should have a dust cover, but this should not restrict through ventilation. Doors which have been shrink-wrapped should be left with the wrapping intact until required for conditioning (as explained below at 4.17.2).

4.17.1 Protection of door framework

If polished or painted, jambs and linings are at risk of being damaged by heavy and/or awkward objects (fittings and fitments) being rubbed against them as they are being moved about the building. They should be temporarily clad or boxed (see Chapter 3, Fig. 3.57) to a height of not less than 1 m. Sills to exterior door frames which are continually passed through by building operatives should be securely covered with a detachable board.

4.17.2 Conditioning

Doors should be introduced slowly into the climatic conditions in which they will remain. It is important, therefore, that interior doors are not present in a building which is drying out after the operation of any wet trade (laying concrete screeds or plastering, etc.), because of the risk of high moisture intake into the wood.

4.18 Door hardware (ironmongery)

In order for it to operate correctly, door hardware (often referred to as 'ironmongery') can be divided into four sections in relation to its use:

Hardware	Use
Hinges	Door hanging devices
Locks and latches	Door restraint and security
Closers	Self-closing mechanisms
Door furniture	Handles, knobs, letter-plates, etc.

4.18.1 Hinges

The types of hinges available are varied depending on the type of door to be hung. For a door such as a ledge-braced-and-battened door, tee-hinges (Fig. 4.69) are used. They are formed from light gauge metal and finished in a black epoxy lacquer, often referred to as 'black-japanned', or they may have a galvanized finish for extra protection. Two types are available (Fig. 4.69): the standard type for lightweight doors, or the 'Scotch' hinge, which is more robust for heavier doors such as garage doors. These are screwed to the face of the door and frame, as the thickness of the door is insufficient to receive a standard type hinge.

Thicker doors require standard cranked type hinges, which are housed into the edge of the door and the inside edge of the frame and are listed below.

A standard cranked steel butt hinge is shown in Fig. 4.70. The knuckle of the hinge must protrude beyond the face of the door leaf and the edge of the door jamb. Use 100 mm hinges on exterior doors and 75 mm ones on interior doors.

Loose-pin butt hinge (Fig. 4.71): once the door leaf has been hung, it may be detached from its frame by removing the pins from the hinge knuckle – this is ideal for light-duty door sets.

Lift-off butt hinge or loose butts (Fig. 4.72): a medium-duty hinge which enables the door leaf to be detached from its frame.

Cast butt hinge (Fig. 4.73a): made from cast-iron, this hinge is very hardwearing and is classed as 'heavy-duty'.

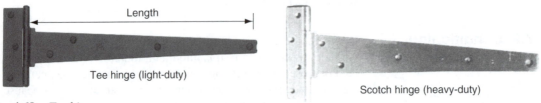

Tee hinge (light-duty)

Scotch hinge (heavy-duty)

Length

Fig. 4.69 *Tee hinges*

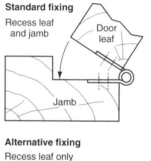

Standard fixing
Recess leaf and jamb

Door leaf

Jamb

Alternative fixing
Recess leaf only

Door leaf

Jamb

Standard cranked steel butt

Fig. 4.70   *Standard cranked steel butt hinge*

Fig. 4.71   *Narrow-pattern butt hinge with loose cup-head pin*

Heavy duty stainless steel hinge with bearings (Fig. 4.73b): this type of hinge is used in areas where there is likely to be above-normal usage. For example, main entrance doors, etc.

Brass butts with phosphor bronze washers (Fig. 4.73c): again used in areas of high use. The phosphor washers reduce the amount of ware on the brass knuckles.

Hurlinge (Fig. 4.74): this light-duty butt hinge does not require housing into the door or jamb; provided the correct screw size is used, the leaf thickness of the hinge provides door clearance. Care must be taken not to permit a build-up of paint around the hinge when painting, otherwise a condition known as 'hinge bind' will result – this is when, on

Standard fixing

Recess leaf and jamb
(Anticlockwise closing shown)

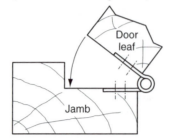

Alternative fixing

Recess leaf only
(Anticlockwise closing shown)

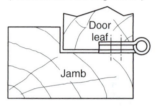

Fig. 4.72 *Lift-off butt hinge or loose butts*

closing the door, the hinge leaves touch before the door leaf is fully closed.

Rising or skew butt hinge (Fig. 4.75a): this is a lift-off hinge with a helical knuckle which causes the door to rise as it is opened. This movement enables the bottom of the door to clear carpets, etc. and to fall shut under the door's own weight, thereby providing a simple self-closing device.

Because the door starts to rise as soon as it is opened, the top inside corner (hinged side) of the door will need to be slightly splayed, as shown in Fig. 4.75b, to avoid catching the door head or transom rail.

Parliament hinge (Fig. 4.75c): a hinge that is used on a door frame situated in a deep wall reveal. If a convential hinge were to be used, the door would bind against the wall preventing it to be opened more than 90°. A parliament hinge will allow the door to swing on a wide radius, thus clearing the wall (Fig. 4.75d).

The disadvantage of this particular hinge is the projecting knuckles when the door is closed. This may not only be unsighty but can 'catch' a person's clothes whilst walking by.

Standard fixing

Recess leaf and jamb

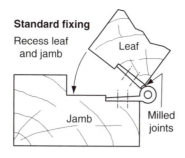

Leaf

Jamb

Milled joints

Alternative fixing

Recess leaf only

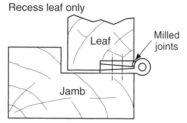

Leaf

Milled joints

Jamb

(a)

Note: *Care must be taken when fitting these hinges – although they are very strong they are also quite brittle and can be damaged if struck with a hammer.*

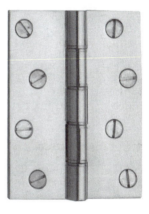

(b)

(c)

Fig. 4.73 *(a) Heavy duty cast iron butts; (b) Double ball bearings stainless steel hinge; (c) Brass butt with phosphur bronze double washers and brass pins*

Standard fixing

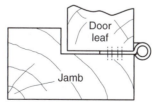

Door leaf

Jamb

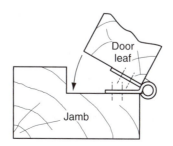

Door leaf

Jamb

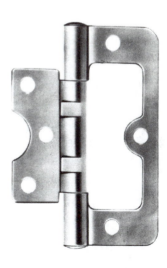

Fig. 4.74 *'Hurlinge' hinge (note: available with loose cup-head pin)*

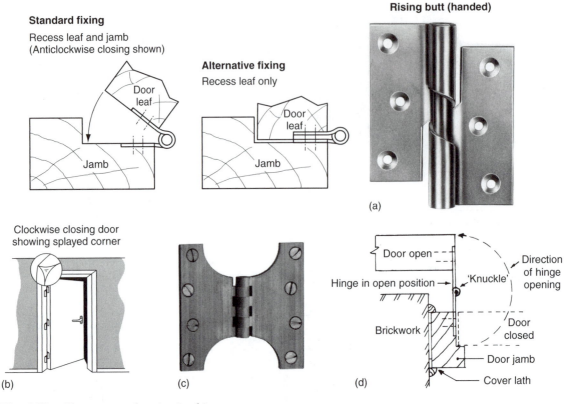

Fig. 4.75 *Clearance and projection hinges*

Hinge bolts (Fig. 4.76): The use of hinge bolts will prevent the door from being forced of its hinges – particularly in outward opening doors where intruders may try to remove the hinge pins to gain entry.

The bolt may take the form of a steel rod attached one hinge leaf (shown in Fig. 4.76a), which engages into a hole in the opposite leaf (attached to the door frame) when in the closed position.

Alternatively, other types include independent bolts (Fig. 4.76b) that are secured both in the edge of the door and frame.

4.18.2 Latches and locks

Once the door has been hung, the door schedule is consulted to determine which type of locking mechanism is to be used depending on the type of door and its location.

Further subdivision will be necessary:

(a) thumb latch (Fig. 4.77);
(b) rim locks (Fig. 4.78b, also see Fig. 12.14a)

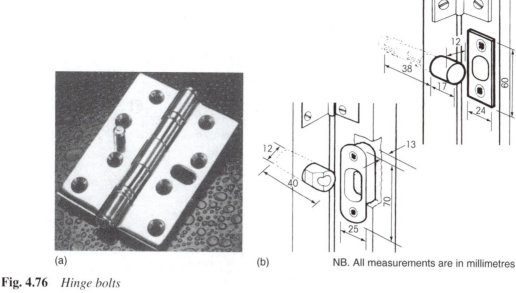

NB. All measurements are in millimetres

Fig. 4.76 *Hinge bolts*

(c) cylinder mortise locks (Fig. 4.79f);
(d) mortise locks (Fig. 4.79);
(e) mortise latches (Fig. 4.80);
(f) cylinder rim locks (cylinder rim nightlatches) (Fig. 4.81).

4.18.2.1 Thumb latch (ledged-braced-and-battened door) (Fig. 4.78) (also see Fig. 4.2)

The thumb latch, sometimes known as a Norfolk or Suffolk latch, provides a simple yet trouble-free means of holding a ledge-braced-and-battened door closed whilst at the same time allowing it to be opened from both sides. If, on the other hand, the door has to be fixed shut at some time for security reasons, then the addition of a lock and/or bolt should be considered.

It should be fixed in the following manner:

1. Make a slot in the door, into which the sneck is pushed;
2. Screw the face-plate (to which the sneck is hinged) to one face of the door;
3. Fix the beam and keeper to the opposite side;
4. The stop can now be screwed to the frame.

The ledged-braced-and-battened door, because of it thickness compared to standard doors, requires hardware that is fixed onto the face.

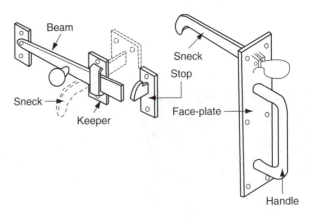

Fig. 4.77   *Thumb latch (Norfolk or Suffolk latch)*

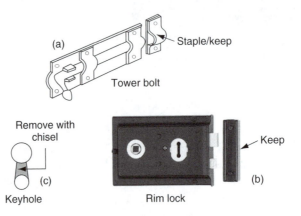

Fig. 4.78   *Locks for ledged-braced-and-battened door*

Probably the simplest way of making the door secure is by the use of a tower bolt, screwed to the door's ledge (see Fig. 4.78a).

To lock the door, a rim lock or rim dead lock (Fig. 12.14a) is used in areas of security (Fig. 4.78b), which will require a keyhole to be made in the door (Fig. 4.78c).

4.18.2.2 Lock/latch terminology

Backset (Fig. 4.82)   The distance by which the vertical centre line of the keyhole and/or spindle hole is set back from the front of the outer forend.

Blank   An unnotched key – see 'Key blank'.

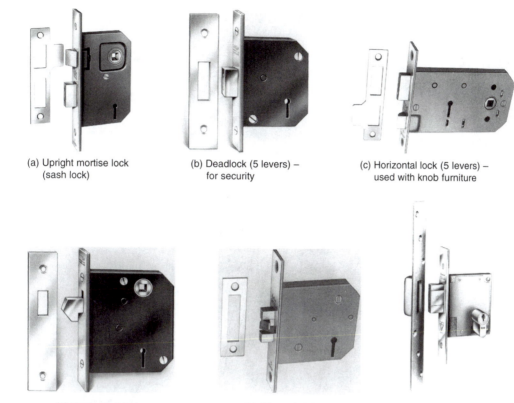

(a) Upright mortise lock (sash lock)

(b) Deadlock (5 levers) – for security

(c) Horizontal lock (5 levers) – used with knob furniture

(d) Hook – bolt lock (for sliding doors)

(e) Claw – bolt lock (for sliding doors)

(f) Cylinder dead lock

Fig. 4.79 *Mortise locks*

Bolt (Fig. 4.82) The part of the lock or latch which protrudes from the case and enters a striking-plate or keep to lock or latch the door.

Case (Fig. 4.82) Contains the lock and latch mechanism.

Claw bolt (Fig. 4.79e) Sections of this bolt move sideways, dovetail fashion, as the bolt is shot (pushed out of its case via a key). It is used on sliding doors.

Cylinder (Fig. 4.84) This houses the pin or disc tumbler and spring mechanism. It may be round-faced or profiled.

Cylinder lock or latch (Figs 4.79f) A lock or latch which uses a cylinder key mechanism (also available as an upright mortise lock (sash lock)).

Deadbolt (Fig. 4.82) A square-ended bolt moved by a key.

Deadlatch (Fig. 4.81) A nightlatch with a spring bolt capable of being locked with a key or catch.

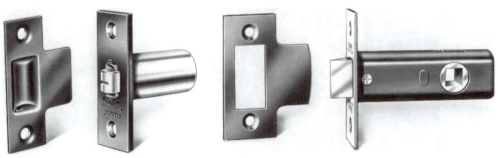

Tubular mortise latch (roller latch)   Tubular mortise latch

General-purpose mortise latch

Fig. 4.80 *Mortise latches*

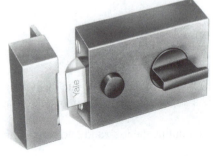

Double-locking nightlatch (spring-bolt   Nightlatch (narrow stiles)
can be deadlocked) (wide stiles)

Twin-cylinder automatic deadlock

Fig. 4.81 *Cylinder rim locks (cylinder rim nightlatches)*

Deadlock   (Fig. 4.79b) A lock with a square-ended bolt operated by a key or thumb-turn.

Disc tumbler lock   A cylinder lock with discs instead of pin tumblers (not dealt with in this chapter).

Double-handed lock   A lock which, by being turned upside down, can be used on either clockwise or anticlockwise closing doors. The keyway has a hole at both ends of the slot.

Escutcheon (plate)   A cover for a keyhole – part of the lock furniture.

Face-plate   (the outer of a double forend) A cover plate fixed to the inner forend.

Follower   The part of the bolt mechanism which withdraws the spring-bolt as it is turned. It may have one or more horns (see Fig. 4.83).

Forend   (Fig. 4.82) Mortise lock or latch fixing plate.

Full-rebated forend   (Fig. 4.85) A purpose-made forend or attachment specially shaped for use either with a rebate in a door jamb or (more likely) with door-to-door meeting stiles.

Handing   (Fig. 4.53) With locks and latches, interior or exterior operation may also be necessary.

Hook bolt   (Fig. 4.79d) Used for securing sliding doors – as the bolt (which is pivoted) enters the keep, it lifts up and then drops over the keep.

Horizontal lock   (Fig. 4.79c) This allows a greater distance between the door edge and the spindle hole or keyhole. It is used with knob furniture.

Keep or keeper   Fixed on or into the door jamb to retain the bolt (Fig. 4.78b).

Key   See Fig. 4.83d.

Key blank   A key sized to fit a keyhole, but not notched to suit the locking mechanism.

Latch   (Figs 4.80 and 4.81) A bevelled spring bolt, or a roller bolt used to fasten a door. Certain types are lockable, e.g. nightlatches (Fig. 4.81).

Levers   (Figs 4.83a and b) Flat shaped plates which form part of the locking mechanism – the greater the number, the better the lock's security. Figure 4.83a shows their operation; Fig. 4.83b shows how they are fitted within the lock case.

Lever handles   (Fig. 4.91) Lock or latch furniture which operates the spring bolt of a lock or latch.

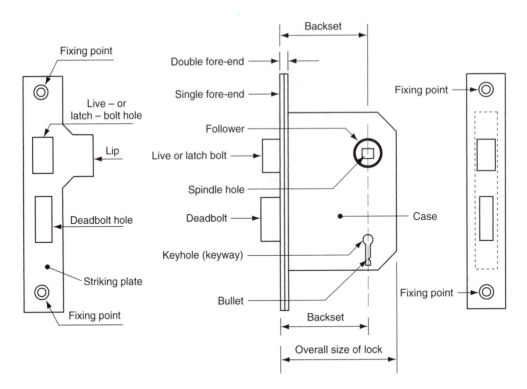

Fig. 4.82 *Mortise-lock components*

Lock (Figs 4.79 and 4.83b) A case containing a locking device, usually operated by a key.

Locking latch (Fig. 4.81) A latch (spring bolt or roller bolt) capable of being locked.

Lock set Matched lock and furniture (handles, spindle, escutcheon, etc.) ready for fixing.

Mortise lock or latch (Figs 4.79 and 4.80) A locking or latching device which is mortised into a door stile.

Narrow-case lock or latch A rim lock or latch suitable for narrow door stiles.

Nightlatch (Fig. 4.81) A latch with a bevelled spring bolt or a roller bolt, which automatically latches as the door is closed. The latch may be withdrawn from its keep by a key from the outside or by a knob or lever from the inside. Latches may also be retained within their case, or deadlocked.

Pin tumbler mechanism (Fig. 4.84) A cylinder mechanism consisting of a row of spring-loaded tumblers which rise and fall to meet

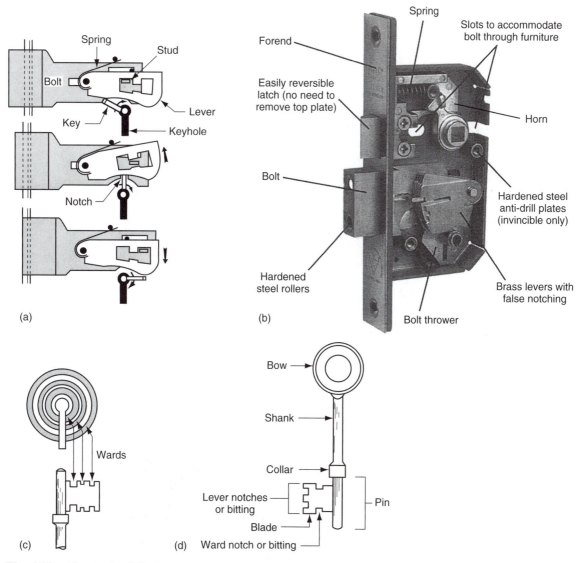

Fig. 4.83 *Mortise latch/lock mechanisms*

the V-notches cut in the key. When all the junctions between pins (the lower parts of the tumblers) and the drivers (the upper parts of the tumblers) are aligned (by using a correctly cut key), the cylinder plug can be rotated to operate the latch.

Plug (Fig. 4.84) The part of the cylinder into which the key is put. It can be rotated when the correct key is used to allow the pins and drivers to separate. It may also have a disc tumbler mechanism (not dealt with in this chapter).

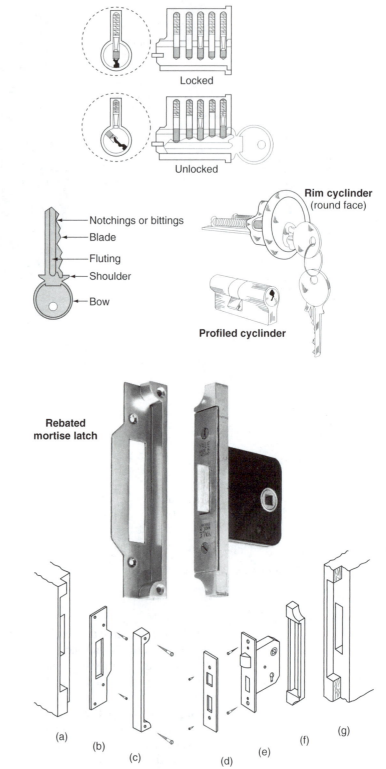

Fig. 4.84 *Cylinders and their operation*

Fig. 4.85 *Full-rebate components sets: (a) rebate in door jamb, (b) striking-plate, (c) rim section, (d) plain forend; (e) mortise lock, (f) rebated forend, (g) rebate in door*

Rebated lock or latch (Fig. 4.85)   A mortise lock or latch with a full-rebated or adapted forend.

Reversible bolt   A spring bolt which can be turned round to suit the door handling. *Caution* The bolt is under pressure from a compressed spring, and this spring can become dislodged if the case is opened. If it is necessary to open the case, wear eye protection and carry out the operation within a tall-sided open-top cardboard box.

Rim lock or latch   A lock or latch which is screwed on to the face of a door.

Roller bolt   (Fig. 4.80) An alternative to a bevelled spring bolt – it does not require handing.

Rose   A plate which fits behind a cylinder or a door knob.

Sash lock   (Fig. 4.79a) The same as an upright mortise lock.

Sash wards   See 'Wards'.

Shoot of a bolt   The distance a bolt moves.

Skeleton key   A key cut to suit a number of differently warded locks, by bypassing the wards. (This is not possible with a cylinder or levered lock.)

Spindle   A square-sectioned steel bar which forms part of the lock furniture. It passes through the follower and fits into a lever handle or knob on each side.

Spring bolt   (latch bolt) (Fig. 4.82) A bolt which is spring-loaded to retain it in the extended position. One edge is bevelled to form a lead to enable it to ride over its striking-plate before entering its keep. It is operated by handles or knobs.

Spring-latch   (Fig. 4.81) A latch with one spring bolt, operated by a key from outside and by a knob from inside.

Staple   A box-shape keep fixed to a door jamb to receive the bolt of a rim lock or latch.

Striking-plate   (Fig. 4.82) A metal plate let into and screwed to the door jamb to shroud the mortise holes which receive bolts from mortise locks and latches.

Tubular mortise latch (catch)   (Fig. 4.80) A mortise latch which fits into a single round hole.

Upright lock   A mortise or rim lock with a narrow case to fit into (sash lock) or on to the width of a door stile.

Wards   (Fig. 4.83c) Metal rings fixed to the inside of the lock case such that a key must pass around them in order to operate the bolt – often used with levers.

Warded lock   A lock which relies solely on wards as a means of security – suitable only where security is of little importance.

4.18.2.3 *Fitting and fixing a mortise lock*

(i) Determine the position (usually 900 m/m) of the spindle hole. For asthectic reasons, locks fitted to multi-light doors should have their spindle hole in line with the glazing bar nearest to the middle of the door.

(ii) At a convenient open position, gently wedge the door to prevent it swinging. (Overtightening will cause damage to the door and frame, and possibly to the wall.)

(iii) Lightly mark the top and bottoms of the mortise hole (Fig. 4.86a) and transfer across the door edge (Fig. 4.86b).

(iv) With a marking gauge or combination square, gauge the centre of the door leaf and the amount of backset (Fig. 4.86c).

(v) Measure and transfer to the door the distances down the backset line to the centres of the spindle hole and keyhole (Fig. 4.86d).

(vi) Bore the spindle hole and the keyhole as shown in Figs 4.86e and f – from both sides of the door leaf.

(vii) With a chisel, cut out the keyhole from both sides of the door leaf (Fig. 4.86g).

(viii) Using a bit with a diameter equal to the width of the lock case, and a depth stop, bore a series of holes for the mortise (Fig. 4.86h).

(ix) Chop and clean out the mortise hole square – if a swan-necked mortise chisel is available it is most useful for this purpose (Fig. 4.86i).

(x) Position the lock in the hole. Mark around its forend (Fig. 4.86j), then chop out a recess in the door's edge until the forend fits flush. Screw the lock to the door.

(xi) Fit the spindle and handles to the door. Check that the spring bolt and key operate.

(xii) Close the door. Mark the top and bottom of the latch bolt and dead-bolt on to the face of the door jamb (Fig. 4.86k).

(xiii) Square across the rebate. Gauge distance 'x' (Fig. 4.86l). Chop out the mortises in the jamb to form the keep.

(xiv) Adjust the mortise holes in the door jamb until the latch bolt and dead bolt operate freely, then position the striking-plate over the holes. Mark out the recess and chop out until the striking-plate is flush with the woodwork – screw back to the door jamb.

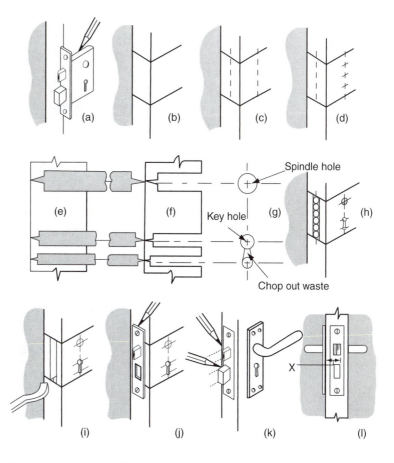

Fig. 4.86 *Fitting and fixing a mortise lock*

4.18.3 Door closers

Door closers are mechanical devices which force or encourage a door to close on its own. Some 'push', others 'pull', and rising butts (already described under the heading 'Hinges') simply rely on the weight of the door leaf. Methods of application vary – some are attached to the door and/or the frame, some form part of the hinge mechanism, while others are concealed.

Figure 4.87 shows examples of an overhead door closer that can be used on either hand of door, and can be adapted to suit most hanging situations. It contains a hydraulic check, adjusted by a turn-screw, which prevents the door from slamming. Manufacturers fixing template is also supplied with the unit.

The closer shown in Fig. 4.88 fixes on to the edge of the jamb and is suited for doors up to 50 kg. The tension arm slides across a nylon guide strip as the door is opened, and closes under pressure from a coil spring in the tube. Pressure adjustment is made with a detachable tommy (tension) bar, as shown in the diagram. This is screwed into the closer and, in this case, rotated to the right to increase the tension. The arm is then screwed into the left corresponding hole and, when located back on the door, the tommy bar is removed. This is a simple yet functional mechanism.

Door closer fitted to door with arm fixed to face of frame head

Door closer fitted to door with arm fixed to underside of frame head

Door closer fitted to face of frame with arm fixed to door

Fig. 4.87 *'Briton' overhead door closer – three methods of application*

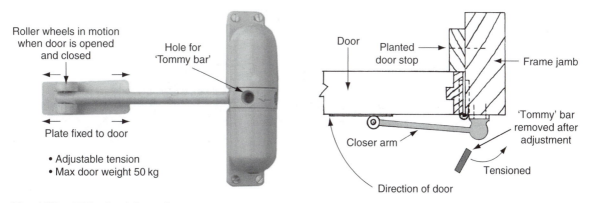

Fig. 4.88   *'Gibralter' door closer*

Figure 4.89 shows a concealed door spring which fits inside the edge of the door and is attached via a chain to the door jamb. The mechanism is only visible when the door is open.

A 'dictator' door check is shown in Fig. 4.90a, which may be fitted to the top of the door (other applications are available including horizontal fixing). This is generally used with door closers shown in Figs 4.88 and 4.89 to eliminate the 'slamming' action of the door, which may lead to the frame becoming loose and the hinges being damaged.

The hook is secured to the head of the frame to receive the wheel of the 'dictator'. As this wheel engages, hydraulic fluid contained in the cylinder passes through a valve – this will dampen or slow down the door as it engages into the hook. The 'dictator' door check is ideal for situations where the door needs to be securely locked shut, for example entrance doors to offices or flats.

When a pair of doors with rebated closing stiles are used, together they will be required to close in the correct sequence. For this reason, a door selector can be used (Fig. 4.90b). This ensures that the door with the leading edge will always end up in its correct final position. Again with the assistance of the door closers shown in Figs 4.87, 4.88 and 4.89.

4.18.4 Door furniture

Door furniture consists of those items of hardware which are fixed either directly or indirectly to the face of a door and/or its framework. Manufacturers often divide such items into the following groups:

(a) Lever furniture (Fig. 4.91)

(b) Knob furniture (Fig. 4.92)

} or { Lock and/or latch furniture; or Mortise-lock and rim lock furniture;

(c) Letter-plates, postal knockers (Figs 4.93 and 4.97c);

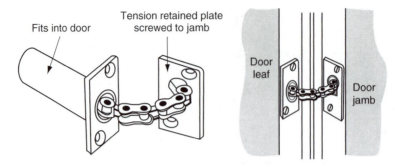

Fig. 4.89 *'Chasmood adjunct' door closer*

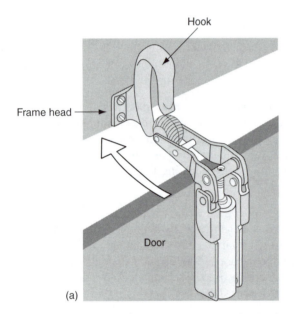
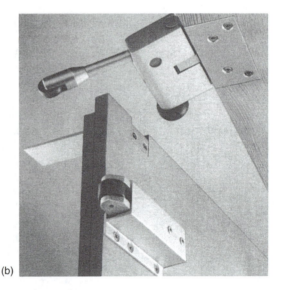

Fig. 4.90 *Door checks. (a) Dictator door check; (b) 'Union' door selector*

(d) Finger and kicking plates (Fig. 4.94);
(e) Door handles and pulls (Fig. 4.95);
(f) Security furniture (Fig. 4.96);
(g) Sundry items of furniture (Fig. 4.97).

Figure 4.97 gives examples of sundry items of door furniture:

(a) Proprietary door stops are used to protect the wall and to prevent straining the hinges if opened too far.
(b) Shows one type of door coat hanger, but this may also make up a number of hangers on a board fitted to the door or wall.
(c) Shows a letter plate with anti-draught flap, similar to Fig. 4.93.
(d) Shows two types of bell push: the conventional battery/transformer type, and the other a solar powered. The ring box may be located anywhere within the dwelling without the need for wiring.

Interior door
Exterior door

Fig. 4.91 *Lever furniture*

Latch lever handle
Lock lever handle

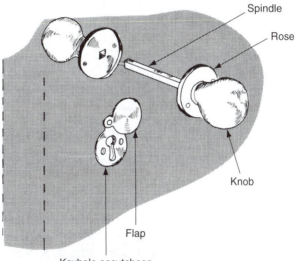

Spindle

Rose

Knob

Flap

Keyhole escutcheon

Fig. 4.92 *Knob furniture*

Fig. 4.93 *Combined letter-plate/postal knocker*

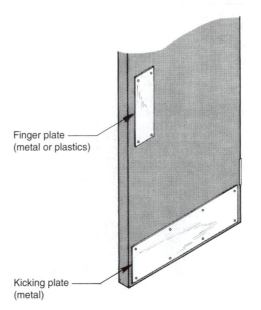

Fig. 4.94 *Finger and kicking plates*

Finger plate
(metal or plastics)

Kicking plate
(metal)

Fig. 4.95 *Door handles and pulls*

Cylinder pull
(for night cylinder latch)

Bar pull

Flush pull
(for sliding doors)

Barrel bolt

12mm
Body dia

25 mm to
55 mm

Door viewer

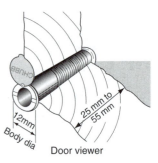

Fig. 4.96 *Security furniture*

Door chain (alarm chains available)

Indicator bolt

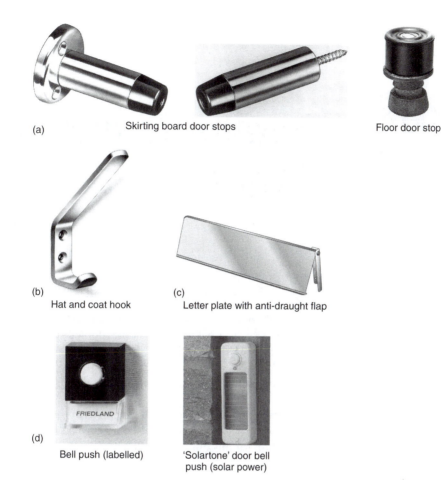

(a)　Skirting board door stops　　　　　　Floor door stop

(b)　Hat and coat hook

(c)　Letter plate with anti-draught flap

(d)　Bell push (labelled)　　　'Solartone' door bell
push (solar power)

Fig. 4.97 *Sundry items of door furniture*

References

BS 459: 1988, Specification for matchboarded wooden door leaves for external use.

BS 6206: 1991, Impact performance.

BS 6262–4:1994, Glazing for buildings. Safety related to human impact.

BS 8214: 1990, Code of practice for fire door assemblies with non-metallic leaves.

BS 5588–0:1996, Fire precautions in the design, construction and use of buildings. Guide to fire safety codes of practice for particular premises/applications.

BS 6206: 1981, Specification for impact performance requirements for flat safety glass and safety plastics for use in buildings.

BS 6262: 1982, Code of practice for glazing for buildings.

BS 544: 1969, Specification for linseed oil putty for use in wooden frames.

BS 952–1:1995, Glass for glazing. Classification PAS 23.

Building Regulations Approved Document B, Fire Safety: 2000.

5

Domestic garage doors

Garage doors are much the same in their construction as some exterior house doors (see Chapter 4). The main door openings are of course wider, and doors are therefore hung in pairs, unless a one-piece door is used, in which case it would be of the 'up-and-over' type or made to slide. (Large sliding and folding-leaf doors are outside the scope of this book.) See Work Activities – by Brian Porter & Reg Rose Chapter 6. Table 5.1 shows common sizes that are available.

Because these doors are made to open outwards and are often left open for long periods, provision should be made in their design and construction for all-round protection against the weather.

5.1 Garage door types

Only three types/systems are to be considered here:

(i) ledged-braced and battened double-leaf doors;
(ii) framed-ledged braced and battened double-leaf doors;
(iii) 'up-and-over' single-door systems.

5.1.1 Ledged-and-braced battened double-leaf doors (Fig. 5.1)

Methods of construction and hardware for these doors are shown in Chapter 4. Each door is side-hung either with Scotch-tee hinges (see Fig. 4.69) or with bands and hooks (crooks) (see Fig. 5.3), which offer greater security and longer service. General security can be provided

Table 5.1 Garage doors (nominal door opening size)

Height	Width	Thickness
Double-leaf side-hung wood garage doors		
6'6" (1981)	7'0" (2134)	1¾" (44)
7'0" (2134)	7'0" (2134)	1¾" (44)

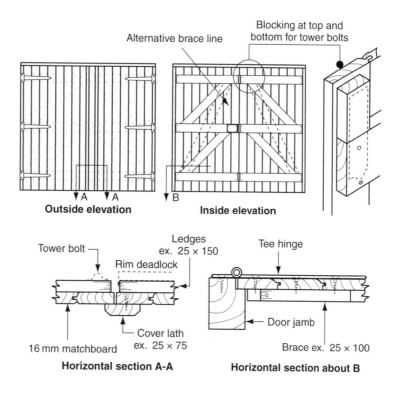

Fig. 5.1 *Ledged-and-braced battened double-leaf doors*

by two tower bolts (see Fig. 5.4) – one at the top, the other at the bottom of the inside face of the door – and a rim deadlock.

If, due to the door's width, brace lines (sag bars) are at an angle of less than 45° to the horizontal, then a two-piece (via the middle ledge) single brace may be adopted as shown.

Where doors meet, a cover lath is fixed (screwed from the inside of the door) to the door leaf which is to receive the lock. This forms a door-handing rebate.

5.1.2 Framed-ledged braced and battened double-leaf doors (Fig. 5.2)

Construction details are shown in Fig. 5.2. These are heavier and stronger doors than those shown in Fig. 5.1 and usually require stronger hinges, i.e. bands and hooks. Glazed panels may be included in the design as shown in Fig. 5.2. Meeting stiles can be rebated from the solid, or the doors may be reduced in width to accept two planted staff beads (the former method is better where security is of special importance).

5.2 Fitting and hanging hinged doors

Before starting, check the door opening for plumb and square-ness and that it coincides with the door sizes (use a lath or pinch rod).

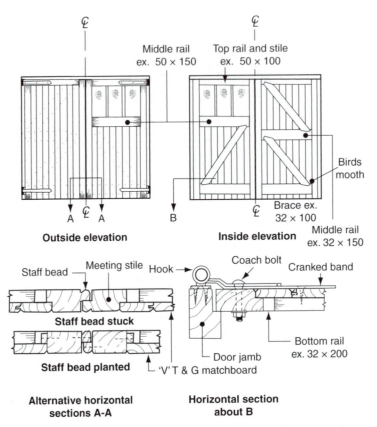

Middle rail
ex. 50 × 150

Top rail and stile
ex. 50 × 100

Birds mooth

Brace ex.
32 × 100

Middle rail
ex. 32 × 150

Outside elevation

A   A

B

Inside elevation

Staff bead   Meeting stile   Hook

Coach bolt   Cranked band

Staff bead stuck

Staff bead planted

Door jamb

'V' T & G matchboard

Bottom rail
ex. 32 × 200

**Alternative horizontal
sections A-A**

**Horizontal section
about B**

Fig. 5.2   *Framed-ledged braced and battened double doors – with or without glazed units*

5.2.1 Fitting doors (one at a time)

(i)   Remove the top and bottom horns;

(ii) Fit the hanging stile – scribe if necessary;

(iii) Provide top and bottom door clearance – 3 mm top clearance and 10 mm floor clearance. Scribe the bottom of the door to the floor if necessary;

(iv) Provide an edge clearance of not less than 3 mm between the hanging stile and the jamb, and 3 mm between meeting stiles (excluding any leading edge).

5.2.2 Hanging door

(i)   With each door laid flat across two battens supported on trestles, position a band (with its hook in place so that door jamb position can be allowed for) on to each of the top and bottom rails. Temporarily fix with two screws per band.

(ii) Position both doors into the opening (provide a check (restraint) around the framework). Pack and adjust for correct clearance. Tack a lath across the front of the doors and jambs to retain clearance. Brace as necessary to restrain the doors (this is very important when assistance is not available and/or on windy days).

(iii) Slide the pin of each hook into the band loops. Temporarily fix with two screws per hook.

Note: Because each pin is a loose fit within its loop, the top hooks should be lightly pushed away from the door as they are being fixed; conversely, the bottom hooks should be pushed towards the door. In this way, and provided the shoulders of the pins were held tightly up against the bottom of each band loop, the door should not drop or fall out of plumb when the door packings are eased and then removed together with the front distance lath.

(iv) After removing the packings, etc., lift the doors off their hinge pins (hooks). Lay them flat once more and, dealing with one band at a time, mark the position of the coach bolt (a bradawl may be used for this purpose) then remove the band. Bore a hole to receive the bolt, paint the back of the band and its door position, then fully fix the band back on to the door, including the coach bolt, not forgetting to use a steel washer behind the nut (see Volume 1, pages 167–8). This procedure should be repeated with all the bands.

(v) Adopt a similar procedure with each hook but, before removing it from the jamb, gently tap the base of each pin (collar) so that an impression is left on the jamb where the pin was welded to the back-plate. Any protruding bits of metal can be let into the jamb by cutting a shallow sloping groove with a chisel – alternatively a hole may be bored, provided it slopes upward to prevent water from lodging behind the crook. Only two screws per hook should be used at this stage.

Note: It is important that, after the bands and hooks are removed at stages (iv) and (v), they are returned to their original positions. If, as mentioned, each one is dealt with separately, mistakes will be avoided.

(vi) Seal the bottom edge of the doors by painting and/or treating with preservative. Particular attention should be paid to the end grain of the stiles.

(vii) Rehang the door. Check for all-round clearance and for twist. Small amounts of twist may be removed by recessing the back-plate of the offending hook or hooks (diagonally opposite) into the jambs.

(viii) Fit and fix stop laths and a threshold as necessary.

(ix) Fix remaining hardware – security and stays, etc.

(x) Glaze as necessary.

5.3 Hardware

Hardware for side-hung garage doors can be divided into the following:

(a) hinges;
(b) security;
(c) door stays.

5.3.1 Hinges

Tee hinges are dealt with in Chapter 4, section 4.18.1.

A cranked band and hook is shown in Fig. 5.3. The crank (bend) in the band (strap) enables the face of the door to be in line with the front edge of its jamb. When the door is in the open position, it can be lifted off its hinges – this useful feature permits easy access for maintenance; for example, all future painting can be done under cover.

Added security can be achieved by replacing the screws which attach the hooks to the jambs with clutch screws.

Fig. 5.3   *Cranked band and hook*

Fig. 5.4   *Tower bolt*

Fig. 5.5   *Garage door bolt (adjustable 'drop' bolt)*

Fig. 5.6   *Bow or D-handle bolt*

Fig. 5.7   *Monkey-tailed bolt*

Fig. 5.8 *Foot bolt*

Fig. 5.9 *Hasp and staple*

5.3.2 Security (bolts)

One door leaf will be held shut by bolts. The most common type of bolt, and the least expensive, is the tower bolt (Fig. 5.4). Figure 5.5 shows a much larger version known as an adjustable 'drop' bolt.

Stronger and heavier bolts such as the 'bow' or 'D-handle' bolt (Fig. 5.6) and the 'monkey-tail' bolt (Fig. 5.7) generally have handles within easy reach of the operator, thereby avoiding any over-reaching or bending. These heavy-duty bolts are square in section and spring-loaded to prevent them dropping when used vertically at the top of the door.

An alternative bolt for the bottom of the door, which is also operated without bending, is the foot bolt (Fig. 5.8). As its name implies, this is shot by pushing down with the foot. The bolt is retracted by pressing down on the side pedal.

The other door leaf can be held closed by a rim deadlock (see Fig. 12.14a) or a padlock with a hasp and staple. The type shown in Fig. 5.9 has a floating staple to compensate for any moisture movement, and all its fixing-screw heads are concealed when fastened.

5.3.3 Stays

Stays provide a means by which garage doors may be safely held open (at at least 90° to their opening). The door already fitted with a bolt may be provided with an extra keep for when the door is open, and a further bolt may be added to the foot of the adjacent door. If the doors hinge back to a wall or fence, probably the simplest and most effective means of providing a stay is by using two cabin hooks (Fig. 5.10).

Fig. 5.10   *Cabin hook and staple*   For fixing detail see Fig. 6.13

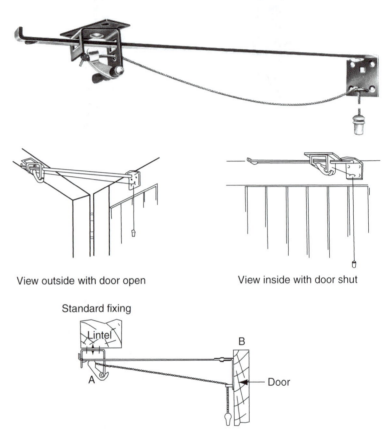

View outside with door open   View inside with door shut

Fig. 5.11   *Garage door holder*

Failing that, the most satisfactory method – shown in Fig. 5.11 – is to use two Crompton garage door holders; their automatic check secures the doors in the open position. The doors are closed by simply releasing the catch by pulling on the acorn pendants, then closing the doors by hand. *At no time should the cords be used to close the doors.*

5.4 'Up-and-over' single-door systems

These doors are more likely to be made from galvanized steel, aluminium, or GRP (glass reinforced polyester) than timber, but the great advantage that the timber door has over the others is that purpose-made sizes are easily constructed. Generally accepted door-opening sizes for these one-piece doors are shown in Table 5.2.

Table 5.2   Door-opening sizes for single
leaf up-and-over garage doors

Width	Height
6′6″ (1982)	6′6″ (1982)
6′6″ (1982)*	7′0″ (2134)
7′0″ (2134)	6′6″ (1981)
7′0″ (2134)*	7′0″ (2134)
7′6″ (2286)	6′6″ (1982)
7′6″ (2286)	7′0″ (2134)
8′0″ (2438)	6′6″ (1982)
8′0″ (2438)	7′0″ (2134)
9′0″ (2743)	6′6″ (1981)
9′0″ (2743)	7′0″ (2134)
14′00″ (4265)	6′6″ (1981)

* Most common sizes

Figure 5.12 shows two possible types of door. Fig. 5.12a uses a framed-ledged-and-braced battened method of construction, whereas Fig. 5.12b has a solid-timber outer frame with a central muntin and midrails. For the sake of appearance, the outer surfaces of the two exterior-grade plywood panels have been subdivided by two false muntins.

When these doors are fully open (overhead), their own dead weight makes them inclined to sag across their width. This deflection can be overcome by fixing a purpose-made brace longitudinally along each rail. These braces may be available from the door-gear manufacturers (Fig. 5.12i), but possible alternatives are shown in Fig. 5.12ii – iv.

5.4.1 Door-gear

'Up-and-over' gear must match the door's weight and material and may be fitted by a number of methods including:

  (i) Canopy;
 (ii) Tracked;
(iii) Optimizer.

Figure 5.13 shows the principle by which these door systems operate:

  (i) *Canopy* (Fig. 5.13a): a common system suitable for most wood doors, which when in the open position leave one-third of the door protruding (providing a canopy).
 (ii) *Tracked* (Fig. 5.13b): includes horizontal tracks secured to the wall or ceiling to support and allow the door to retract completely back into the garage.

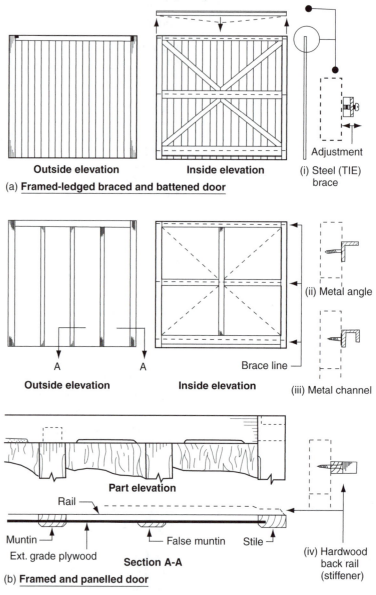

In the horizontal position, deflection will occur due to the dead weight of the door, restraint can be offered by using cross-members

Outside elevation   **Inside elevation**   (i) Steel (TIE) brace

(a) **Framed-ledged braced and battened door**

A   A

Outside elevation   **Inside elevation**   (iii) Metal channel

(ii) Metal angle

Brace line

Part elevation

Rail

Muntin   False muntin   Stile

Ext. grade plywood   **Section A-A**

(iv) Hardwood back rail (stiffener)

(b) **Framed and panelled door**

Fig. 5.12   *'Up-and-over' single-leaf garage doors*

Note: *Unless ordered separately, hardware for up-and-over doors comes complete with the operating gear – see manufacturers' catalogues.*

(iii) *Optimizer* (Fig. 5.13c): the lifting gear is secured to the door frame with the tracks on the door – which eliminates the need for horizontal tracks.

Figures 5.13b and c are recommended for double size doors of timber, metal or GRP, with the addition of remote control. Sliding and folding door gear is shown in Work Activities, B. Porter and R. Rose (2004) Elsevier, Chapter 6.

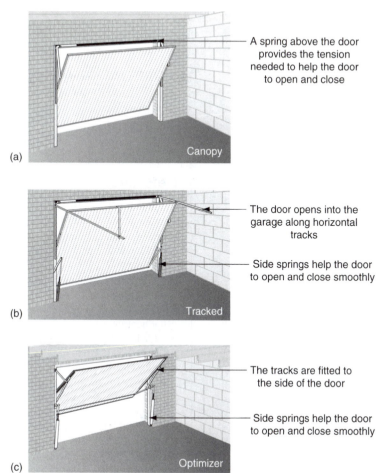

A spring above the door provides the tension needed to help the door to open and close

(a)

The door opens into the garage along horizontal tracks

Side springs help the door to open and close smoothly

(b)

The tracks are fitted to the side of the door

Side springs help the door to open and close smoothly

(c)

Fig. 5.13 *'Henderson' operating gear for 'up-and-over' doors*

5.5 Surface finish

Note: *See Section 2.6 regarding high/medium and low finishes.*

Note: *Micro-porous stains may contain fungicides. They can also offer resistance to ultra-violet rays that otherwise could lead to the eventual breakdown of their protection (see vol. 1, Chapter 2)*

Sliding garage doors are usually purchased 'self finished', which means they have been coated in a suitable protective coating against the elements. Depending on the material of the door, this coating is built up in layers and may be classified as low to high build as seen in the following examples:

(i) Timber

Paint – exterior paint (medium/high build) includes primer, undercoat and finish coat in matt or gloss.

Stain – either spirit, water or oil based (low build) and available as a semi-transparent coloured coating applied to hardwoods or softwoods to enhance the grain of the timber.

Varnishes – (low build) include syntectic, copal and water thinned – contain PVA which is low odour and quick drying.

(ii) Steel

Usually galvanized and finished with coating of polyester powder (high build) which has been baked on during the manufacturing process.

(iii) GRP (glass reinforced polyester)

Includes a smooth gloss or simulated timber finish which is almost maintenance free.

References

BS 459: 1988, Specification for matchboarded wooden door leaves for external use.

BS 6262: 1982, Code of practice for glazing for buildings.

BS 544: 1969, Specification for linseed oil putty for use in wooden frames.

BS 952–1: 1995, Glass for glazing. Classification.

Building Regulations Approved Document B, Fire Safety: 2000.

BS 1282: 1999, Wood Preservatives. Guidance on choice use and application.

BS 4261: 1999, Wood Preservation, Vocabulary.

BS 5589: 1989, Code of practice for wood preservation.

BS 5268–5: 1989, Structural use of timber, code of practice for the preservative treatment of structural timber.

BS 5707: 1997, Specification for preparations of wood preservatives in organic solvents.

BS 7956: 2000, Specification for primers for woodwork.

BS 6150: 2005, Code of practice for painting of buildings.

BS 7664: 2000, Specification for undercoat and finishing paints.

BS EN ISO 4618–3: 2000, Paints and varnishes. Terms and definitions for coating materials. Surface preparation and methods of application.

BS EN 12051: 2000, Building hardware.

BRE Digest 301: 1985, Corrosion of wood by metals.

6   Domestic gates

Gates are means by which persons and/or vehicles can pass through what could otherwise be a physical barrier to a property or area. The degree of security offered by a gate will depend on:

(a) its type;
(b) its method of construction;
(c) its height;
(d) its ability to be locked.

The importance of the above will greatly depend on the purpose of the gate. For example, the security of a garden, driveway, side, or yard gate for a domestic dwelling will eventually be at the discretion of the house owner; on the other hand, commercial and industrial premises will require their gates to be heavily secured as and when required. These latter types of gate are outside the scope of this book. However, they may resemble hinged solid garage type doors shown in Chapter 5.

6.1 Single personnel-access and garden gates

These gates are not usually wider than 1 m and, unless a visual barrier is required, are generally slatted so that wind pressure is less of a problem and water is driven off their surfaces more quickly.

Figure 6.1 shows three examples of slatted gates. All have been designed in such a way that their upper surfaces are either sloping or covered by a weathered capping to discourage water from lying upon them. The greatest danger from wet rot is where palings lap rails and braces and abut ledges or rails. Particular attention to pre-painting/preserving is important around these areas.

As shown in Fig. 6.1 gates (a) and (b) can be made up of a mortised-and-tenoned framework. Joints should be either wedged and glued (with exterior-type resin adhesives) or painted and pegged (dowelled). Alternatively, a ledged-braced-and-paled construction could be used. Gate (c) must be framed, braced, and paled. Gaps between palings should be narrow enough to prevent a toe-hold, thereby preventing

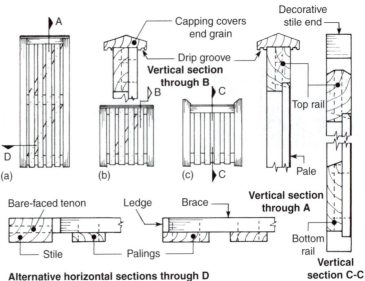

Fig. 6.1 *Single personnel-access and garden gates*

Alternative horizontal sections through D

rails being used as a ladder and discouraging children from swinging on the gate.

6.2 Domestic vehicle- and personnel-access gates

Where the gateway serves only vehicles, it is worth considering a wide single gate. On the other hand, if it is to double as a garden gate, two (double) gates would be more practicable – regularly opening and closing a wide gate can be awkward and difficult.

Figures 6.2 and 6.3b show a modified field gate. Its design lends itself to having a personnel gate to open across one-third of the overall opening width to give greater ease of access. However, such arrangements are not always practicable with openings wider than 2.4 m, as the gate stop and/or bolt keep could interfere with vehicle tyres.

Figure 6.3a shows one of a pair of traditional-type double garden gates. Both gates are capable of being swung, but one is usually kept ground-bolted until vehicle access is required. The bolt (see Figs 5.4 to 5.8) is shot into the driveway via a protruding timber, metal, or concrete gate stop, which can be a hazard when both gates are open.

To give the appearance of a traditional field gate, an extended cruck (bent) stile has been used. In Fig. 6.3c and d this serves as a straining anchor for the through or cross-tie; but in Fig. 6.3b, although functional, it is mainly ornamental.

Timber for the cruck stile would have originally been taken from naturally bent wood. Today, this design is copied by laminating timber to whatever shape is required.

Fig. 6.2 *Modified field gate to include personnel access*

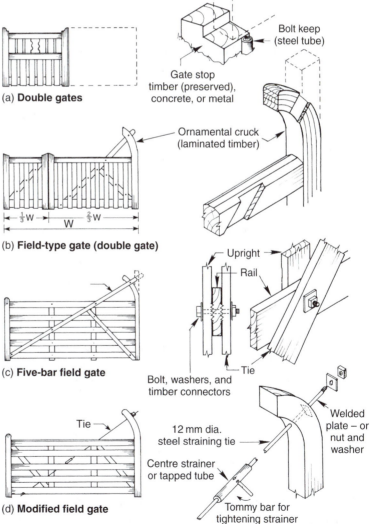

(a) **Double gates**

Bolt keep (steel tube)

Gate stop timber (preserved), concrete, or metal

Ornamental cruck (laminated timber)

(b) **Field-type gate (double gate)**

$\frac{1}{3}$W   W $\frac{2}{3}$W

(c) **Five-bar field gate**

Upright

Rail

Tie

Bolt, washers, and timber connectors

(d) **Modified field gate**

Tie

12 mm dia. steel straining tie

Centre strainer or tapped tube

Tommy bar for tightening strainer

Welded plate – or nut and washer

Fig. 6.3 *Domestic vehicle- and personnel-access gates*

Meeting stiles for the gates should be dealt with as shown in Fig. 5.2 (garage doors).

Figure 6.3c shows a five-bar field gate of simple construction. The top rail and stiles – one of which is extended, either straight or as a laminated cruck – are made from 75 mm thick timber. Mid-rails (mortised into the stiles), the upright tie, and the brace can be bolted together with timber connectors (Volume 2, p. 145, Fig. 8.21) where they intersect.

Figure 6.3d shows a fully framed five-bar gate with a tie made from end-tapped steel rod. The rod passes through the gate and is anchored by a plate and nut under the bottom rail. Tension adjustment can be made via the plate and nut sited on the hanging edge of the cruck. Alternatively, tension can be achieved via a mid-strainer made from a piece of internally tapped steel tube.

6.3 Gate hanging

Gates can be hung from brick walls or pillars (Fig. 6.6), concrete posts (Fig. 6.7), or timber posts (Fig. 6.8). Each should be plumb and well anchored to the ground or substructure.

The procedure for hanging gates differs from that already mentioned for garage doors only in the amount of clearance left. Between the gate and its post, 12 mm clearance is allowed for face-hung gates – the gap for gates hung on hooks between posts will depend on the hook protrusion. Floor clearance may be as much as 75 mm.

6.3.1 Sloping paths

If a driveway or path slopes towards the swing of the gate or gates, there are three methods of dealing with the situation, depending on the amount of slope:

(i) use double gates with sufficient gap under the gates to clear the ground when fully open (Fig. 6.4a),
(ii) level sufficient of the driveway to allow the gates to open fully (Fig. 6.4b),
(iii) use purpose-made offset (cranked) bands and hooks at the bottom hinge positions – these will enable the gate or gates to tilt upwards as they are opened (Fig. 6.4c).

If the outside pavement is sloping and the driveway is in line with it, as shown in Fig. 6.5, one end of the gateway will be higher than the other. Provided the rails of the gate are level, the stiles and palings may be extended accordingly (see Volume 2, p. 20, re datum lines.)

Note: *Where hinges are face-fixed, a certain amount of lift can be created by chopping the top back-plate of the hook into the post and packing out the bottom; however, gates will not be plumb when they are closed. Similarly, gates hung between posts can have their hooks fixed out of plumb, and in this way the gate can be made to tilt upwards as it opens*

6.3.2 Gate hardware (ironmongery)

In the main made from ferrous metals (containing iron) that are not only subjected to rusting, but may also be in danger from corrosion which can

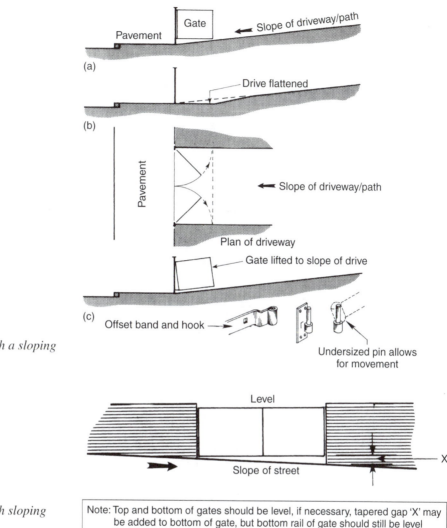

Fig. 6.4 *Dealing with a sloping path or driveway*

Fig. 6.5 *Dealing with sloping street pavements*

be accelerated by the acid condition of certain woods, which becomes active as their moisture content is increased (see Volume 1, page 50(g)).

Under normal conditions ironmongery (to a certain extent) can be protected by painting, but if acidic conditions do exist – such as when using English oak or western Red Cedar – then ironmongery should be galvanized or black-powered coated to prevent direct contact, as painting provides only short-term protection (see Volume 1, Chapter 3).

Hardware falls into the following categories:

- hinges;
- catch or latch mechanism;
- bolts and stays.

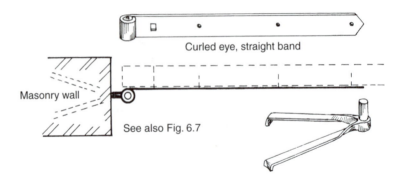

Curled eye, straight band

Masonry wall

See also Fig. 6.7

Fig. 6.6 *Double-pronged hook to brickwork*

6.3.2.1 Hinges

Figure 6.6 shows a double-pronged hook, which is secured in the brickwork to receive a straight (uncranked) band type that is to be fixed to the face of the gate.

> Note: *The hinge 'band', together with its 'hook and plate', is some-times collectively referred to as a 'band and crook'*

Figure 6.7 shows two methods of securing a hook and plate, either direct to the timber post or to the concrete post via a timber facing. Notice that the timber batten has been splay cut on one end to assist with dispersing any surface water.

Heavy gates of wider sectioned timber will require a double-strap (or band and crook) field-gate hinge and hook bolt as shown in Fig. 6.8. (The long band via both sides of the top rail, the short band towards the bottom of the stile.) Both are held in place to the gate via coach bolts and screws. The top hook is bolted through the gate post, as it will be in tension; the bottom hook can be driven into the gate post as this will be in compression. The thickness of the gate fits between the strap and is secured through the gate with coach bolts. The hook may be driven or bolted into the post.

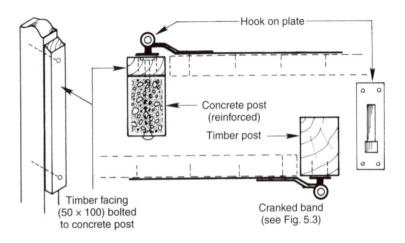

Hook on plate

Concrete post (reinforced)

Timber post

Timber facing (50 × 100) bolted to concrete post

Cranked band (see Fig. 5.3)

Fig. 6.7 *Hook and plate to timber gate post or timber facing to concrete gate posts*

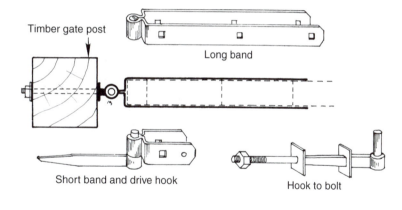

Fig. 6.8 *Double-strap field-gate hinge and hook bolt – for heavy gates*

Timber gate post

Long band

Short band and drive hook

Hook to bolt

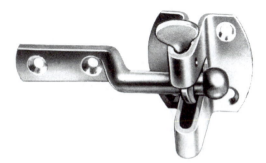

Fig. 6.9 *Automatic gate catch*

6.3.2.2 Catch or latch mechanism

Of the many examples, three are listed below:

- automatic gate catch (Fig. 6.9);
- spring-loaded automatic gate catch (Fig. 6.11);
- throw-over gate loop (Fig. 6.12);
- thumb-latch – see Fig. 4.77 (also known as Norfolk or Suffolk latch).

Automatic gate catch

Probably the most efficient catch for narrow gates is an automatic gate catch like the one shown in Fig. 6.9 – except for the occasional drop of oil on the speck pivot, it is maintenance-free. The malleable iron peg is screwed to the opening gate, and the speck case is screwed either to the gate post (with a single gate) or to the static gate (held closed by a ground bolt) of double gates.

It is important that an automatic gate catch is accompanied by a stop lath (Fig. 6.10) or some other form of gate check, otherwise the peg will become dislodged after a short time.

Tall gates will require a means of latching accessible from both sides. A thumb-latch (Chapter 4, Fig. 4.77) would be ideal for this

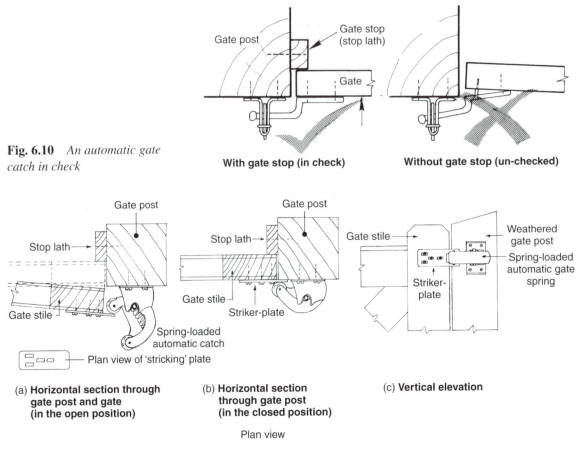

Fig. 6.10 *An automatic gate catch in check*

With gate stop (in check) **Without gate stop (un-checked)**

(a) **Horizontal section through gate post and gate (in the open position)**

(b) **Horizontal section through gate post (in the closed position)**

(c) **Vertical elevation**

Plan view

Fig. 6.11 *Plan view of spring-loaded automatic gate catch*

purpose – possibly with the addition of a horizontally fixed tower bolt (Chapter 5, Fig. 5.4) for added security.

Spring-loaded automatic gate catch

Figure 6.11 shows another type of automatic gate catch that is activated via a spring-loaded rocker arm with rubber rotatable wheels at either end. It is activated when struck by the striker-plate attached to the closing stile – usually in line with the top rail of mid-height gates.

Figure 6.11a shows the rocker arm about to be struck in the open position. Figure 6.11b on the other hand shows the gate closed with the catch in the closed position. Figure 6.11c gives a vertical view behind the gate, with the gate and catch closed.

However, single modified field gates will require a more substantial means of latching to hold the gate closed (see Fig. 6.12).

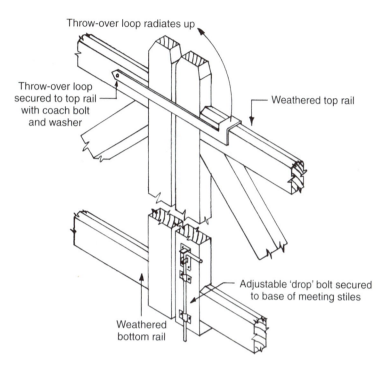

Throw-over loop radiates up

Throw-over loop
secured to top rail
with coach bolt
and washer

Weathered top rail

Adjustable 'drop' bolt secured
to base of meeting stiles

Weathered
bottom rail

Fig. 6.12   *Throw-over gate loop*

Throw-over gate loop (Fig. 6.12)

Where such a gate incorporates a personnel gate (Figs 6.2 and 6.3b), a 'throw-over gate loop' is usually used (see Fig. 6.12). This is bolted to the personnel access gate and loops over the secured field gate. When lifted, it allows the gate to be opened separately, allowing entry.

6.3.2.3 Anchor bolts and stays

There are various methods used to allow the gate to be held either in the open or closed position.
For example:

- Tower bolt (see Figs 5.4 and 4.78);
- Adjustable 'drop' bolt (Figs 6.12 and 5.5);
- Cabin hook and staple (Figs 6.13 and 5.10);
- Gate holder (Fig. 6.14);
- Gate spring (Fig. 6.15).

Tower bolt

Secured vertically or horizontally to the stile of a medium sized gate, its purpose is to hold it in either the open or closed position (see Fig. 5.4).

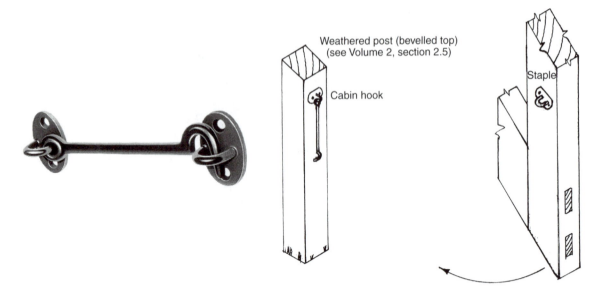

Fig. 6.13 *Cabin hook and staple*

Adjustable 'drop' bolt (Fig. 6.12)

The temporary static gate (depending on which way the opening gate is handed [right or left-hand opening]) is usually held in the closed position with an anchor bolt like those featured in Figs 6.12, 5.4 and 5.5. Additionally, in some cases these bolts can also be used to restrain the gate in the open position. They are secured onto the stile of the gate (at the base) and locate into the ground (see Fig. 6.3a).

Cabin hook and staple (Fig. 6.13)

Additionally, a cabin hook and staple may be used to hold the gate open. The hook is fixed to the wall or weathered post with the staple secured to the gate stile, or visa versa, as shown in Fig. 6.13 (see also Fig. 5.10).

Gate holder

Figure 6.14 shows a type of gate holder that consists of a bar with a tapered hook and counter-balanced weight.

When the gate is opened, and passes over the tapered hook (which is always in the 'up' position because of the weight), it forces it to go down until the gate locates into the recess, which then holds it in place. To close the gate, pressure is applied (by foot) onto the hook and the gate pulled forward. The tapered hook then returns to the 'up' position ready for the next time the gate needs to be secured.

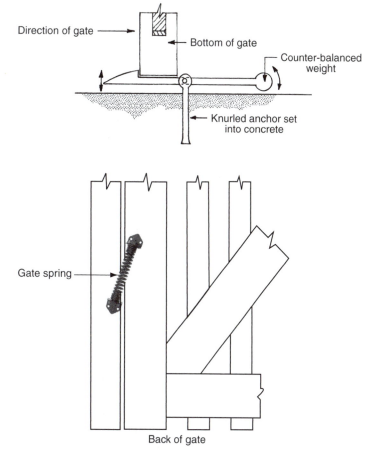

Fig. 6.14 *Gate holder at base of gate*

Fig. 6.15 *Gate spring*

Note: Caution: *Manufacturers' instructions must be carefully read and understood. When charged (wound) these springs are under a great deal of tension and therefore consideration must be given to the health and safety not only of oneself, but also of others who may be around.*

Care must also be taken not to over-tension the spring, which may cause the gate to slam and cause damage or loosen the meeting post.

Gate spring (closer for small to medium weighted gates) (Fig. 6.15)

To ensure that the gate returns to the closed position a coiled gate spring (Fig. 6.15) can be attached to the back of the gate stile (usually about halfway down) and secured to the post with sufficient length screws supplied by the manufacturer.

At each end of the spring a bracket is included (for fixing), which encloses small circular rotating wheels with holes to assist with tensioning.

It is usual for one end of the spring to have a permanent pin in one of the holes to be already in place. Depending on the weight of the gate, the other end of the spring is adjusted (rotated) to the required tension, using a 'tommy bar'. This is a small strong metal pin inserted into one of the small holes in the wheel and rotated in the opposite direction of the gate closing (see also Chapter 4, Fig. 4.88). Once sufficiently tensioned, a smaller, permanent pin is placed into the adjacent hole, and after careful checking to ensure it is in place, the 'tommy bar' can then be removed.

6.4 Suitable material (for gate construction)

Timber is subjected to weather from all sides; therefore, no matter how good the design to reduce end-grain exposure or water traps, unless the timber used is classified as being durable (Volume 1, Tables 1.14 and 1.15), such as English oak, teak, or the softwood western red cedar, or it is suitably treated with a wood preservative (see Volume 1, section 3), the chances are that if surface treatment is not regularly reapplied (be it paint, varnish, or exterior stains) the woodwork will suffer decay – usually from the effects of wet rot (see Volume 1, section 2).

Timber gate posts will of course require the same treatment, particularly at or about ground level (see Volume 2, section 2.5).

References

BS 1282: 1999, Wood Preservatives. Guidance on choice use and application.
BS 4261: 1999, Wood Preservation, Vocabulary.
BS 5589: 1989, Code of practice for wood preservation.
BS 5707: 1997, Specification for preparations of wood preservatives in organic solvents.
BS 6100–1.3.6: 1991, Glossary of building and civil engineering terms.
BS 6150: 2005, Code of practice for painting of buildings.
BS 7956: 2000, Specification for primers for woodwork.
BS 7664: 2000, Specification for undercoat and finishing paints.
BS EN ISO 4618–3: 2000, Paints and varnishes. Terms and definitions for coating materials. Surface preparation and methods of application.
BS EN 12051: 2000, Building hardware.
BRE Digest 301: 1985, Corrosion of wood by metals.

7 Stairs

The primary function of any stair is to provide safe access from one floor, or landing level, to another. The ways by which this is achieved may vary according to the design of the stair or the space available. This chapter as a whole is concerned with the:

(a) design;
(b) construction; and
(c) installation (site fixing) of straight-flight stair to domestic dwellings.

7.1 Terminology

- The term 'stair' refers to a number of 'steps' together with any 'balustrade';
- A 'step' (Fig. 7.1a) consists of one 'tread' and its 'riser';
- A 'flight' (Fig. 7.1b) has a continuous number of steps from a floor or 'landing' (Fig. 7.1c).
- A 'balustrade' (Fig. 7.1d) contains a handrail with an infilling between it and a 'string' (Fig. 7.1e), landing, or floor.

Listed below are some of the terms used in the Building Regulations, Approved Document K: 2000 together with their interpretation.

(a) *Going* (Fig. 7.1f): the horizontal distance across the width of a tread from its nosing to the nosing of the tread or landing above it.
(b) *Landing* (Fig. 7.1c): a floor or platform at the top or bottom of a flight or between flights.
(c) *Nosing* (Fig. 7.1g): the front edge of a tread and/or landing – usually rounded or splayed.
(d) *Pitch* (Fig. 7.1h): the slope of the stair (measured in degrees).
(e) *Pitch line* (Fig. 7.1i): an imaginary line connecting the nosing of each tread within a flight to the nosing of a landing at its top.
(f) *Rise* (Fig. 7.1j): the vertical distance between the upper surface of a tread or landing and that of the tread or landing immediately above or below it.
(g) *Tread* (Fig. 7.1k): the top surface of a step. (It should be noted that this term is also listed below under 'Stair components'.) Other types of treads are shown below.

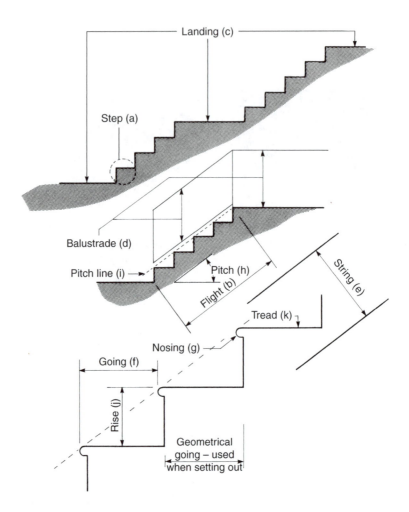

Fig. 7.1 *Stair terminology*

(i) *Alternating tread* (Fig. 7.28) – used on specially designed stairs to fit in an area that would be otherwise unsuitable for a conventional stair.

(ii) *Kite winder* (Figs 7.25 and 7.26) – the centre tapered tread in a set of winders (often referred to as a 'Kite' winder) that fits into the 90° corner radiating out from the newel.

(iii) *Tapered tread* (Figs 7.25 and 7.26) – usually the treads either side of a stair with kite winders in a 90° or 180° turn radiating out from the newel. However, they may be part of a 'spiral' stair (not included in this volume) where all the treads may be tapered.

7.1.1 Stair components

Only those members associated with a straight flight as used in domestic dwellings are listed below. Reference should also be made to BS 585: Part 1, 1989 and BS 5395: Part 1, 2000.

(a) *Apron lining* (Fig. 7.2a): a board used to line the edges of the stairwell.

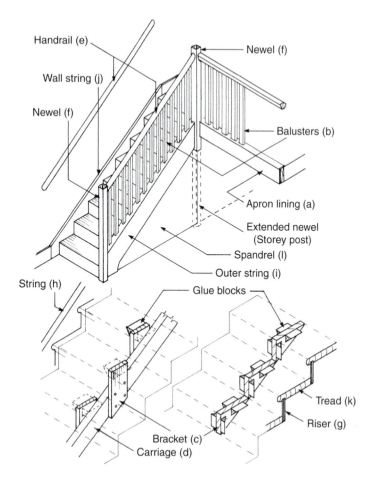

Fig. 7.2   *Stair components*

(b) *Baluster* (Fig. 7.2b): a vertical member forming part of a balustrade.

(c) *Balustrade*: an assembly made up of a handrail with the infilling between the handrail and a floor, landing, or stair string (see section 7.4).

(d) *Bracket* (Fig. 7.2c): either an upright member fixed to a carriage to add support to a tread, or a triangular member (web) fixed to the underside of a tread and the back of a riser.

(e) *Carriage* (Fig. 7.2d): an inclined member to which brackets are fixed upright. It provides extra support to the stair.

(f) *Handrail* (Fig. 7.2e): a handhold, either as part of a balustrade fixed into or on to newels, or fixed separately to a wall above the wall string.

(g) *Newel* (Fig. 7.2f): a post onto or into which a string, handrail, or balustrade rails (Figs 7.8c and 7.29b) are fixed. It forms an end or change in direction of a stair.

(h) *Riser* (Fig. 7.2g): the vertical member of a step.

(i) *String* (Fig. 7.2h): an inclined board into which treads and risers are housed.

Note: *Bottom and top steps may also be fixed to a newel. If the newel post extends from one floor to another it may be termed a 'storey post'*

(j) *Outer string* (Fig. 7.2i): a string away from a wall.
(k) *Wall string* (Fig. 7.2j): a string running against and parallel to a wall – usually attached to it.
(l) *Tread* (Fig. 7.2k): the horizontal member onto which the foot is placed.
(m) *Spandrel frame/panelling* (Fig. 7.2(1)): a panel enclosing members under the slope of an outer string.

7.2 Design

Of the many aspects to be considered when a stair is designed, the following should receive prime consideration:

(a) the siting of the proposed stair – whether there is enough room (this requires accurate site measurement);
(b) construction and installation details must conform with current legislation – Building Regulations, etc.;
(c) suitable material for the stair members;
(d) the finished appearance (aesthetics).

7.2.1 Site details (Figs 7.3 and 7.4)

Unless it only provides access to or from a dais, platform, split-level floor, or balcony, the stair will have to pass through a floor in order to reach its upper landing. The opening in the floor is called a 'stairwell'.

The size (Fig. 7.3), shape, and location of the stairwell will depend on the stair width, pitch, siting, or type. Trimming arrangements to a stairwell are shown in Volume 2, Figs 6.1 to 6.7.

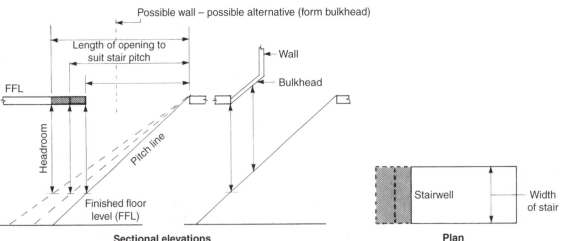

Fig. 7.3 *Factors affecting stairwell*

Stair type is also a major factor which helps to determine the floor space required at the foot of the stair, as shown in the following examples:

(a) Fig. 7.4a – a straight single flight;
(b) Fig. 7.4b – two straight flights with a quarter-space landing between flights turning through 90°;
(c) Fig. 7.4c – two straight flights with a half-space landing between flights, turning through 180° (known as a 'dog-leg' stair);
(d) Fig. 7.4d – three straight flights with two quarter-space landings, turning through 180° (known as an 'open-well' stair).

Figure 7.5 shows where and how measurements are taken for a single straight flight. The 'storey rod' – a stout length of timber (say

> Note: *The flights shown in Figs 7.4b–d need not necessarily be of equal length*

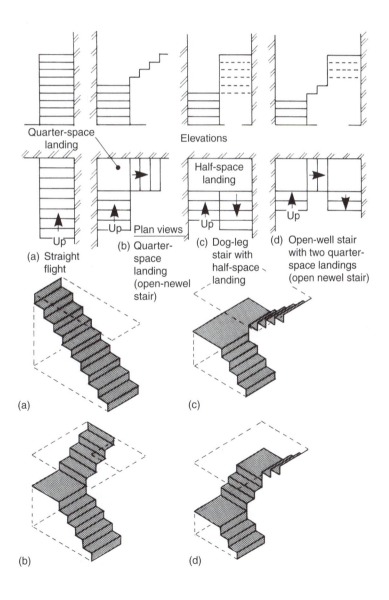

Fig. 7.4 *Examples of floor and well space in relation to stair type*

38 mm × 38 mm section) – must be long enough to reach from the finished floor level (FFL) at (a) to the landing FFL at (b) in Fig. 7.5. Some means of plumbing the storey rod and determining the floor levels must be employed – a spirit level could serve this purpose.

In cases where the FFL has yet to be laid, the thickness of subsequent floor screeds, etc. must be allowed for, otherwise the bottom and/or top rise would not match the others in the flight – any deviation in rise within a flight would contravene the Building Regulations, thereby making the whole stair totally unacceptable.

Measurements and recordable site observations to an existing stairwell would include the following:

  (i)   storey height (total flight rise) – FFL to FFL;
 (ii)   stairwell dimensions – length, width, and depth of floor;
(iii)   stairwell – check for squareness in plan; (Fig. 7.5(c))
 (iv)   any bulkhead projections – trims, etc.; (Fig. 7.5(d))
  (v)   condition of floors, materials used, whether level – if not, by how much out;
 (vi)   total available going – any opening (door or window) or obstructions.

Note: *Openings may be trimmed in a variety of ways – see Volume 2, Figs 6.1 to 6.7*

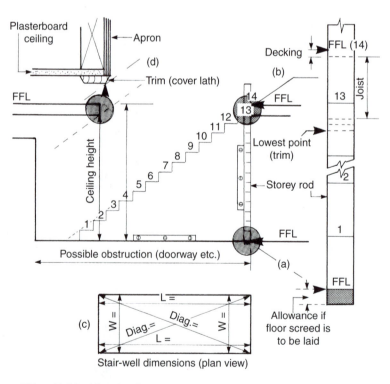

Fig. 7.5 *Taking site measurements for a straight flight and an existing stairwell*

FFL = Finished floor level

If, on the other hand, the stair is to be part of a new development, the stairwell would be constructed in accordance with the chosen stair design.

7.2.2 Building Regulations

Table 7.1 lists the requirements for stairs with straight flights and their landings in accordance with the current Building Regulations (Approved Document K1, Protection from falling, collision and impact), and should be used in conjunction with Fig. 7.6.

An example of how a stair's rise and going (as shown in Table 7.1) can be calculated is described in section 7.8, Stair calculations.

Any doors, including cupboard or duct doors, which open out near the base or landing of a stair (Fig.7.7), must leave a minimum space of 400 mm. This ensures a safe means of escape in the event of an emergency.

7.2.3 Other requirements

Figure 7.8 shows extra considerations based on BS 5395:part 1:2000.

To ensure adequate headroom clearance over stairs with a pitch of $41\frac{1}{2}°$ to $42°$ (the maximum for a private stair), a minimum clearance of 1.5 m measured at right angles to the stair's pitch line should be allowed beneath any ceiling or bulkhead as shown in Fig. 7.8a. If the flight is short enough for a person to jump down the stair, minimum clearance should be increased to 1.8 m (Fig. 7.8b). With an open-riser stair, gaps between treads and risers should be deep enough for the toe of a person ascending or twisting about the stair not to become trapped (Fig. 7.18).

Figure 7.8c shows how a balustrade made up of boards running parallel to the stair or landing could provide a foothold for a child to climb. This type of balustrade should therefore not be constructed in buildings which are likely to house or have their stair used in any way by unaccompanied small children.

7.2.4 Suitability of materials

Timber used in the construction of stairs should comply with BS 1186: Part 2, 1988 and BS1186: Part 3, 1990.

Table 7.2 lists the minimum finished thickness of members for stairs with a close string (one which fully houses the ends of the treads (if fitted) and risers. Open-riser stairs (see Fig. 7.18) will require thicker treads and string members. If treads, risers, or strings are made up of more than one piece of solid timber (strings no more than two), no

Table 7.1   A stair must meet the following conditions as shown in Figs 7.6 and 7.7

Item		Figure	Requirements	
			Private – intended to be used for only one dwelling	Institutional and assembly – serving a place where a substantial number of people will gather
1.	Width of stair	Fig 7.6a	No recommendations are given. However designers must consult Approved Documents B, Fire safety and M, Access facilities for disabled people *	
2.	Pitch of flights	(b)	Max. 42°	Max. 32°
3.	Risers per flight	(c)	Min. 3†‡ Max. 16†‡	Min. 3†‡ Max. 16†‡
4.	Rise of step	(b)	Min. 155 mm, Max. 220 mm* used with going (x) or Min. 165 mm, Max. 200 mm* used with going (y)	Min. 135 mm, Max. 180 mm*
5.	Going of step	(b)	(x) Min. 245 mm, Max. 260 mm* (y) Min. 223 mm, Max. 300 mm*	Min. 280 mm, Max. 340 mm*
6.	Sum of twice the rise plus going ($2 \times R + G$)		Min. 550 mm, Max. 700 mm	Min. 550 mm, Max. 700 mm
7.	Gap between treads (open riser)	(d)	Smaller than 100 mm (should not permit a 100 mm dia. Sphere to pass through it)	
8.	Going of landing	(e)	Not less than the width of the stair (Fig. 7.1)	
9.	Headroom over stair and/or landing	(c)	Min. 2 m (measured vertically from the pitch line)	Min. 2 m
10.	Handrail position	(f)	Flights over 600 mm total rise – one each side for stairs over 1 m wide and over; at least on one side for stairs less than 1 m wide	
11.	Height of handrail to flight	(f)	Min. 900 mm	Min. 900 mm
12.	Height of balustrading guarding a flight	(f)	Min. 900 mm	Min. 900 mm
13.	Height of balustrade guarding a landing	(f)	Min. 900 mm	Min. 1100 mm
14.	Gap between balustrade and pitch line	(g)	Max. 50 mm	Max. 50 mm
15.	Gap between balusters members	(g)	Smaller than 100 mm (should not permit a 100 mm dia. Sphere to pass through it)	
16.	Doors opening out onto landings or near stair.	Fig. 7.7	Doors situated over a landing, including cupboard or duct doors, may open out as long as the space between the open door is not less than 400 mm for safe exit as a means of escape*	

* See Building Regulations (Approved Document K)
† BS 5395: Part 1, 2000
‡ With certain exceptions – see BS 5395: Part 1, 2000
‡ See Building Regulations

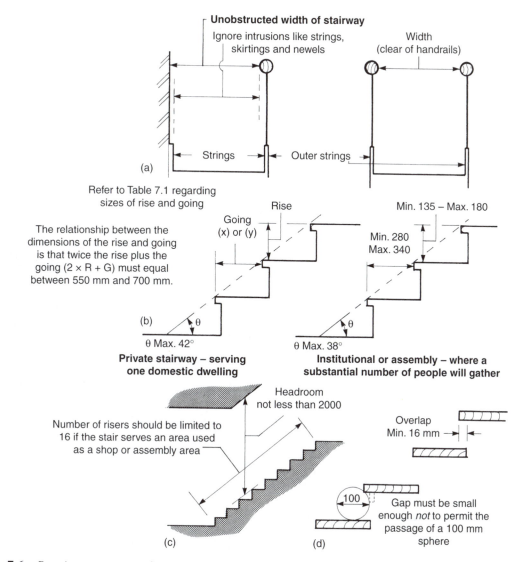

Fig. 7.6 *Requirements to satisfy Building Regulations (Approved Document K) and Table 7.1*

piece should be less than 50 mm wide across its face – fronts of treads should not be less than 90 mm. All joints should be made and glued according to BS 1186: Part 2, 1988. Strings should be at least 225 mm nominal depth.

7.2.5 Surface finish

With a hardwood stair a clear finish is often specified. However, hardwood with a less interesting grain pattern or colour may be stained with a coloured translucent finish. In either case, wood preparation is essential, i.e., all standing procedures must follow the grain.

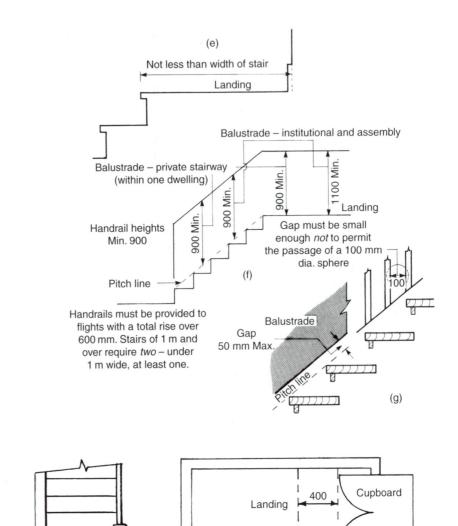

Fig. 7.6   (*Continued*)

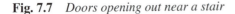

Fig. 7.7   *Doors opening out near a stair*

A softwood stair may also be treated similarly, however, they are more usually painted. Therefore, it has been acknowledged that when sanding, a transversing pattern be employed. This has the effect of levelling the grain, bearing in mind the opaque finish of paint.

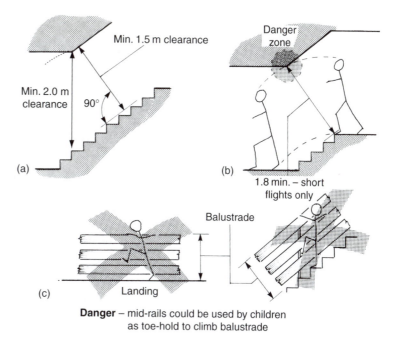

Fig. 7.8 *Additional safety aspects (BS 5395: Part 1, 2000)*

Danger – mid-rails could be used by children as toe-hold to climb balustrade

The important thing is that no matter which finish is used it should be in keeping with its surroundings and all treads must have a non-slip surface.

7.3 Stair construction

Stair construction may be dealt with in the following sequence:

  (i) setting-out;
 (ii) preparation of stair members, i.e. strings, treads, risers (if used), newels, and handrails;
(iii) assembly of stair members;
(iv) bottom step (if applicable);
 (v) balustrades.

7.3.1 Setting-out

This is the most critical part of the job! Any inaccuracies at this stage could result in the overall stair length being too long or too short and once the stair has been assembled very little can be done to rectify any errors. Work should therefore be double-checked throughout all the initial stages of setting-out (marking-off if a storey rod is being used), preferably by the person responsible for taking on-site measurements.

If the stair is to be a 'one-off', it is set out on one of its strings. If, on the other hand, more stairs of the same dimensions (total rise and total going) are to follow – perhaps at a later date – then a workshop rod could be made.

Table 7.2 Minimum finished thickness of stair members (close string and risers)

Method of support	Overall width of stair including string (mm)	Tread thickness (mm)			Riser thickness (mm)			String thickness wood (mm)	Outer newel thickness (mm)
		Wood	Ply	MDF*	Wood	Ply	MDF*		
Supported by a side wall, or up to 9 goings unsupported	Up to and including 990	20	18	**22**	14	9	**12**	26	A square section of 69 or cross section of similar strength and stiffness
	Exceeding 990 but not exceeding 1220	26	24	**25**	14	9	**12**	26	
More than 9 goings unsupported	Up to and including 990	20	18	**22**	14	9	**12**	32	A square section of 90 or cross section of similar strength and stiffness
	Exceeding 990 but not exceeding 1220	32	24	**25**	14	9	**12**	32	

*Recommended sizes only – consult structural engineer.
MDF must be moisture resistant and treads covered with a suitable material, i.e. carpet, to prevent wear.
COSHH regulations must be observed.

Before setting-out can begin a plywood, MDF or hardboard template of the tread and riser end-section – including an allowance for wedging – will be required (Fig. 7.9). In addition to this, a 'pitch board' (Fig. 7.10a) accurately cut to a right angle and with sides equal to the step's rise and going will be required, together with a 'margin template' (Fig. 7.10b) made up of a strip of thin board attached to a timber stock, which allows it to slide parallel with the string for use as a margin width gauge. The margin template also provides a seat for the pitch board, which is used in conjunction with it.

A method of setting out and marking off strings of a short straight flight is listed below and illustrated in Figs 7.11 and 7.12.

(i) *String length* (Fig. 7.11) Provided an allowance is made at each end for the portions above and below the margin line, string length can be calculated using the theorem of Pythagoras (Volume 2, pages 9 and 10). Select the stringboards, mark a face side and true one edge of each board (these must be straight and capable of butting one another squarely), and mark the face edge.

(ii) *Step division* (Fig. 7.12a) Position the margin template stock against the face edge of the wall string (WS) and gauge a margin line down the length of the face side. Then, mark a series of accurately measured steps 'x' along the line according to the number of rises in the flight – in this case five; therefore A to F must equal $5 \times$ 'x'.

(iii) *Transfer* (Fig. 7.12b) Use a try square to transfer the distances 'x' from the margin line to the edge WS. Butt the face edges of WS and the outer string (OS) together – transfer A to F on to OS.

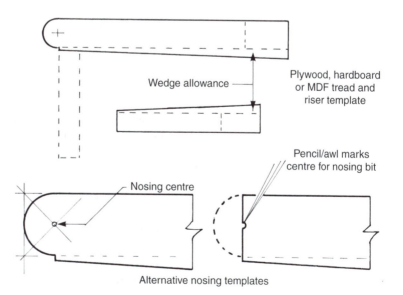

Wedge allowance

Plywood, hardboard or MDF tread and riser template

Pencil/awl marks centre for nosing bit

Nosing centre

Fig. 7.9   *Tread and riser templates*

Alternative nosing templates

Note: If combined boards are used, top and bottom stocks (double stock) must be in line. Independent pitch boards and margin templates (with single stock) are preferred.

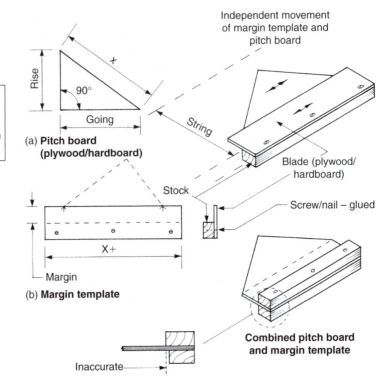

(a) **Pitch board (plywood/hardboard)**

(b) **Margin template**

Fig. 7.10 *Pitch board and margin template*

(iv) *Wall string* (Fig. 7.12c) Working from the left-hand-side face edge (positioned away from you), hold the pitch board against the margin template – as if it were attached to it – and with a pencil mark the FFL and rise '1' from position A to B on the margin line. Slide the assembly between B and C and mark going '1' and rise '2'. Continue like this until the top FFL is reached.

(v) *Outer string* (Fig. 7.12d) Similarly working from the face edge, although this time from right to left and with the pitch board turned over, mark all the rises and goings. Then, using a sliding bevel set to the angle shown, mark both newel positions on the string – riser lines '1' and '5' should fall in the centre of the newels. Using the same bevel settings, mark the extent of the string tenon length 'y' (minimum 50 mm) from the shoulder line as shown in Fig. 7.12e. The ends may now be cut off.

(vi) *Treads and risers* (Fig. 7.12e) Tread and riser templates (positioned according to OS or WS) are laid flat on the string with their face edges against the lines which will form the faces of the treads and risers. (Butting the templates against the pitch board may help here.) In each position, mark around the template until

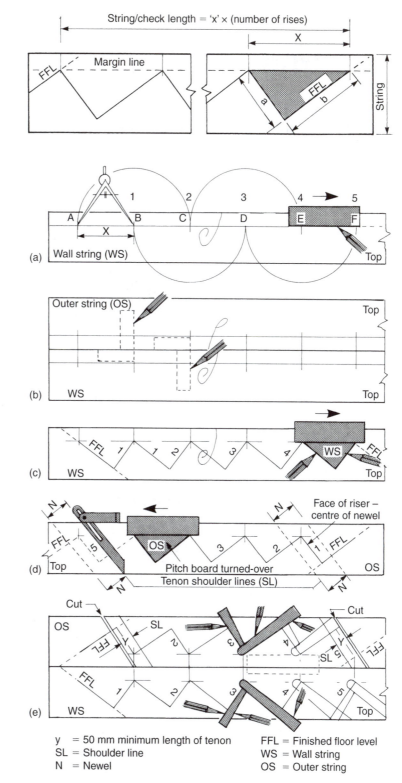

Fig. 7.11 *Determining the true length of the string*

Fig. 7.12 *Setting-out strings*

y   = 50 mm minimum length of tenon   FFL = Finished floor level
SL = Shoulder line   WS = Wall string
N   = Newel   OS = Outer string

> Note: *The same principles apply for setting out longer flights*

the pencil goes off the string – a criss-cross pattern should emerge. These lines mark the extent of the housings that will receive treads and risers. Using a marking gauge, housing depths (minimum 12 mm) are now marked.

7.3.2 Preparation of members

Strings are housed to a depth of not less than 12 mm or 0.4 times the string thickness – whichever is the greater – to receive the ends of the treads and risers (or, in the case of an open-riser stair, the treads – possibly accompanied by a narrow riser, see Fig. 7.18). The method of cutting and shaping the housing with hand tools is shown in Figs 7.13a–f, which illustrate the following stages:

(i) Cut away the nosing portion with a Forstner bit (see Volume 1, Fig. 5.52) with a diameter equal to the tread thickness (Fig. 7.13a). Further holes are then bored as shown. If it is necessary to cut into the riser portion, use the equivalent diameter bit.

(ii) With a firmer chisel and mallet, chop back (slightly undercut) the recessed holes to the riser and tread lines (Fig. 7.13b).

(iii) With the point of a tenon saw in the recess and a small off-cut of hardboard to protect the trench sides, saw (slightly undercut) along the inside face of the riser line down to the depth-gauge line, and again along the wedge-side line (Fig. 7.13c). Repeat the process with the tread housing.

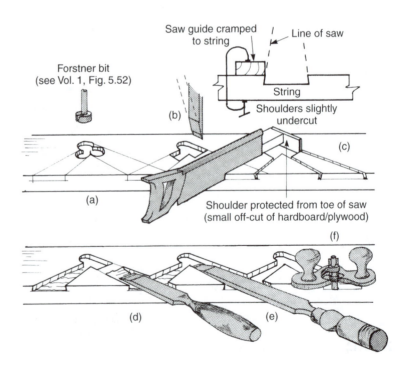

Fig. 7.13 *Trenching a string-board by hand*

(iv) Remove most of the waste wood with a chisel and mallet (Fig. 7.13d) – avoid using the chisel up-turned, as it will tend to dig into the trench bottom.

 (v) Waste wood beyond the reach of a bench chisel can be pared away with a long-bladed paring chisel (Fig. 7.13e).

(vi) Level the bottom of the trenches with a hand router (Fig. 7.13f – see Volume 1, Fig. 5.37).

When a portable electric-powered router is to be used to cut the housings, a jig similar to the one shown in Fig. 7.14a would be suitable for a fixed-base type; but the tread and riser inlet/outlet portions of the jig can be omitted if a plunge-type router is to be used (Fig. 7.14b). (A router template guide and its application are shown in Volume 1, p. 190, Fig. 6.44.)

The string must be firmly held to the bench, just as the jig is to the string (clamps and wedges may be employed), to enable the router to be safely manoeuvred with both hands.

Figure 7.15 shows the use of a proprietary staircase jig clamped to the stair string which is adjustable to suit the required rise and going housings.

Figure 7.16 shows how an outer string is joined to the top and bottom newels with mortise-and-tenon joints. The stub tenons should be not less than 50 mm long and 12 mm thick. If the bottom step extends beyond the newel face (see Figs 7.19 and 7.27), two haunches will be used as shown.

Strings to open-riser stairs will require some provision to prevent spreading. Several of the treads could be tenoned through the strings and/or strings could be screwed to the housed treads. (Screw heads

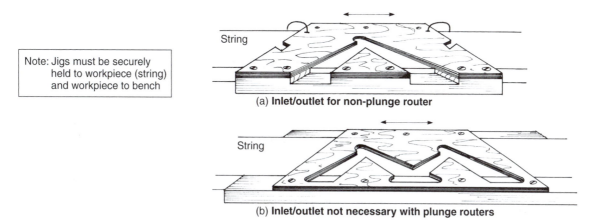

Note: Jigs must be securely held to workpiece (string) and workpiece to bench

String

(a) **Inlet/outlet for non-plunge router**

String

(b) **Inlet/outlet not necessary with plunge routers**

Fig. 7.14 *Jigs for trenching with a portable electric-powered router*

Note: Dust extraction not shown

Clamp

Fig. 7.15 *'Trend' staircase jig*

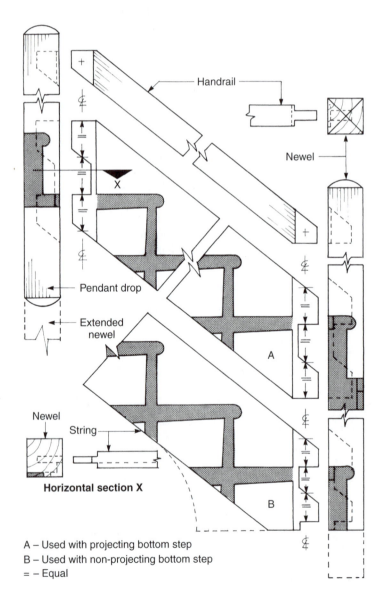

Handrail

Newel

X

Pendant drop

Extended newel

A

Newel

String

Horizontal section X

B

Fig. 7.16 *Outer string and handrail to newels*

A – Used with projecting bottom step
B – Used with non-projecting bottom step
= – Equal

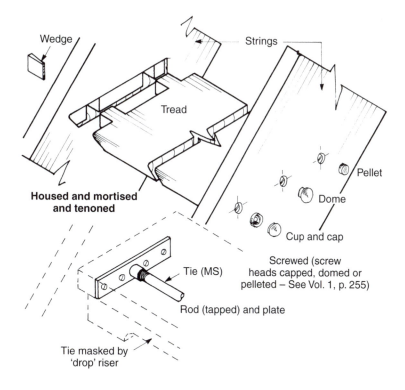

Fig. 7.17 *Preventing string-boards spreading*

Note: *As with the handrails string shoulders may also be housed into the newel posts to conceal any shrinkage*

should be domed or cupped – see Volume 1, page 252, Table 12.2 – or sunk below the surface then covered with a wood pellet (see Fig.7.17).) Steel brackets or tie rods with anchor plates may be other alternatives – see Fig. 7.17.

7.3.3 Treads and risers

Figure 7.18 shows different ways of preparing treads and risers. Treads should be not less than 235 mm wide and their nosings should project not less than 15 mm beyond the face of the risers. (Projection, like tread width, must be uniform throughout the whole flight.) Riser tops should be housed (e.g. plywood risers) or tongued to a depth of not less than 5 mm, but never more than one quarter of the tread thickness.

Riser bottoms may be tongued into the treads or grooved to receive the tread. Open-riser stairs do not have a full riser. However, so that the opening does not contravene the regulations (see Table 7.1), they may use a narrow riser, either hanging down from the underside of a tread (the most common method) or as an up-stand from the back of the tread's surface (this could serve as a dust trap, producing a 'shadow' line with polished stairs, Fig. 7.18).

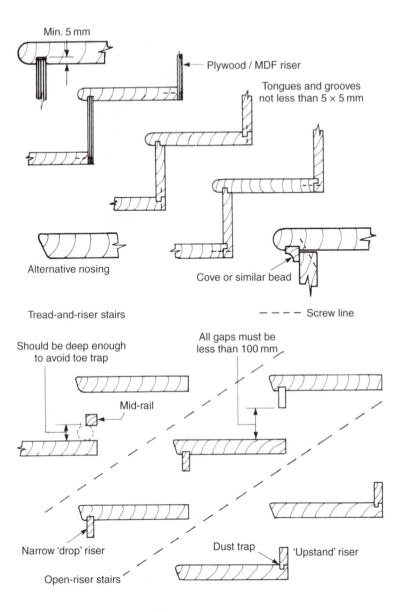

Min. 5 mm

Plywood / MDF riser

Tongues and grooves
not less than 5 × 5 mm

Alternative nosing

Cove or similar bead

Tread-and-riser stairs

– – – – Screw line

Should be deep enough
to avoid toe trap

All gaps must be
less than 100 mm

Mid-rail

Narrow 'drop' riser

Dust trap

'Upstand' riser

Open-riser stairs

Fig. 7.18 *Preparation of treads
and risers*

7.3.4 Newels

Newels will be stop-mortised as shown in Figs 7.16 and 7.19 to a depth of not less than 50 mm. The joints will eventually be tightened by draw-boring and pinning as shown in Chapter 4, Fig. 4.45. Housings which receive treads and risers should be not less than 12 mm deep.

Newel post tops are shaped according to style and ease of handling. The pendant drop (Fig. 7.19) is similarly shaped – elaborate shapes may provide inaccessible dust traps.

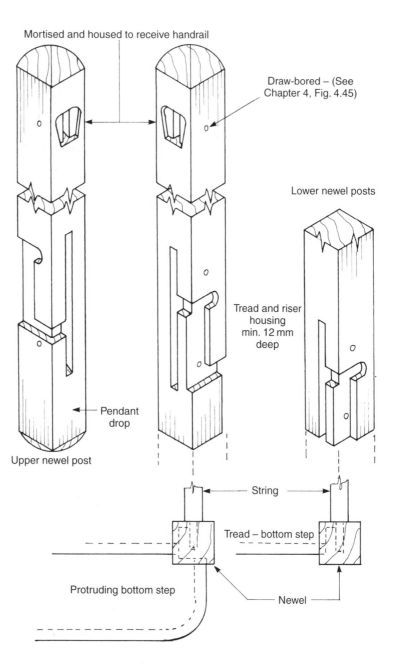

Mortised and housed to receive handrail

Draw-bored – (See Chapter 4, Fig. 4.45)

Lower newel posts

Tread and riser housing min. 12 mm deep

Pendant drop

Upper newel post

String

Tread – bottom step

Protruding bottom step

Newel

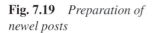

Fig. 7.19 *Preparation of newel posts*

7.3.5 Handrails (Fig. 7.20)

The handrail above a wall string is fixed to the wall either directly (Fig. 7.20a) or via a steel bracket (Fig. 7.20b). A handrail forming part of a balustrade may be fixed to the face of a newel post (Fig. 7.20c). In most other cases, handrails are stub-mortised into the newel post – the underside of some are grooved to receive the tops of balusters as shown in Fig. 7.20d.

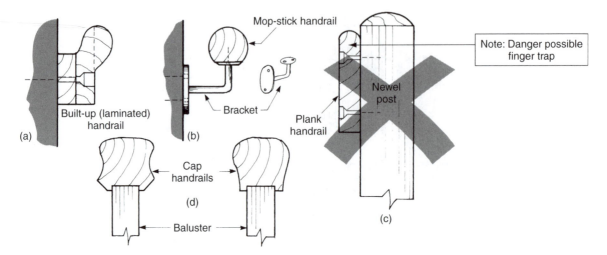

Fig. 7.20   *Examples of handrail sections and their fixing*

7.3.6 Assembly of stair members

Assembly may take place in one of two ways:

- laid flat (Fig. 7.21);
- laid on edge (Fig. 7.23).

7.3.6.1 *Laid flat*

Figure 7.21 shows the various stages involved:

(i) Position strings face edge down on a framework of two runners supported by cross-members (distance pieces). Use blocks to prevent the clamps, which pass under the runners, from falling over. The framework must not be twisted.

(ii) Position, glue, and cramp between strings enough treads (depending on the length of flight) to enable the stair to be checked for square by measuring across the diagonals (see Fig. 7.23(c)), using as measuring points the squaring marks transferred from the face edge of the string at the setting-out or marking-off stage. Use side runners to protect the strings and help distribute cramping pressure.

(iii) With two operatives opposite one another striking in unison, drive glued wedges (ensuring treads are seated tight into the tread housings) between the walls of the string housings and the underside of the treads until tight.

(iv) Fix the remaining treads (except the bottom and top steps if newel posts are to be used) in the same manner – extra cramps may be required.

(v) Cut back protruding wedges to allow the risers to be slid into position. Similarly wedge these risers, then screw them to the back edge of the treads (Fig. 7.22).

(vi) Glue angle blocks (glue blocks) between the treads and risers – two per tread for stairs up to 900 mm wide, three for stairs up to

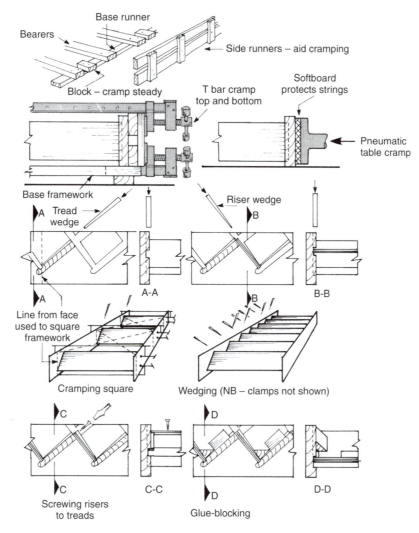

Fig. 7.21   *Stair assembly – laid flat*

990 mm wide, and four for stairs above 990 mm wide (distance between blocks not to exceed 150 mm), as shown in Fig. 7.22.

For stairs with winders (see Fig. 7.25), glue blocks are spaced at 150 to 200 mm from the string and 150 to 225 mm thereafter.

An alternative method of cramping the stair would be by using a proprietary pneumatic cramping table.

7.3.6.2 Laid on edge (Fig. 7.23)

This is generally a popular method of cramping.

(i) Pre-assemble treads and risers into steps with the aid of a jig/cradle (Fig. 7.23a).

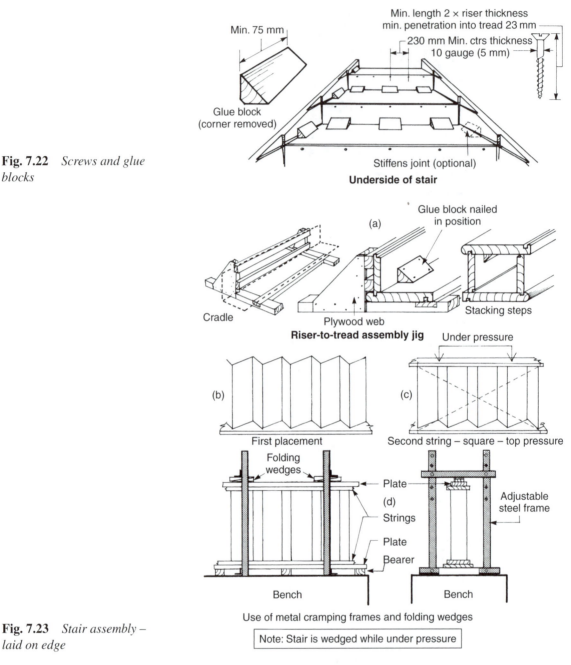

Fig. 7.22 *Screws and glue blocks*

Fig. 7.23 *Stair assembly – laid on edge*

(ii) Lay one string flat on to an assembly bench (with a smooth flat material positioned between the bench top and the string to prevent the string faces being bruised) and position the assembled steps into the string housings (Fig. 7.23b).

(iii) Seat the opposite string (ensuring that treads and risers are squarely positioned in and against the front edge of the string housings), then apply top pressure (Fig. 7.23c) via a purpose-made

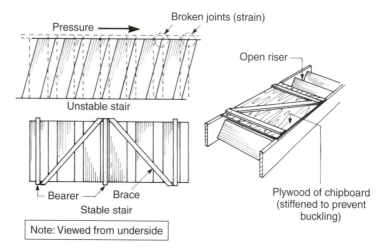

Pressure →

Broken joints (strain)

Open riser

Unstable stair

Bearer — Brace

Stable stair

Plywood of chipboard
(stiffened to prevent
buckling)

Note: Viewed from underside

Fig. 7.24 *Temporary bracing for an open-riser stair*

Note: *If the bottom step is to project forward of the newel, it is built separately and left until the newels have been fixed on site. Similarly, the top riser and nosing may be left loose and fixed on site*

cramping frame which may be fabricated from steel angle, lengths of timber, and hardwood wedges (Fig. 7.23d). In this situation the riser wedges are driven in first to ensure that the Pre-assembled treads and risers are positioned tight into their housings.

(iv) Apply glue to wedges of top and bottom risers first, prior to being driven.

(v) Check for square as previously described at section 7.3.6.1(ii).

(vi) Cut back protruding wedges to allow for tread wedges.

(vii) Continue as described at 7.3.6.1(vi).

Because of the narrowness or absence of risers, an open-riser stair must be handled very carefully until temporary bracing has been fixed as shown in Fig. 7.24, otherwise the joints between the treads and the string may become strained, with the possibility of breaking the glue line.

7.3.7 Straight flights with winders (Fig. 7.25)

Tapered treads or winders are used with a straight flight in most cases (unless it is a spiral stair, see Approved Document K) to provide a continuous turn without the use of landings and to give additional headroom to a stair below as well as providing a pleasing effect to the building.

However, this type of stair may prove to be more expensive to construct and install, including in some cases problems when moving furniture up and down the flights.

Building Regulations Approved Documents, A, D, and K will need to be consulted before this type of stair is to be considered, which does include the following:

(i) tapered treads at their narrowest end (radiating out from the newel) must not be less than 50 mm, see Fig. 7.25f.

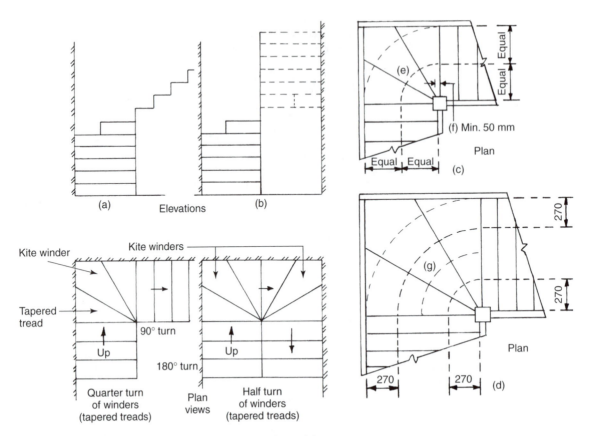

Fig. 7.25 *Samples of stairs with winders (tapered treads)*

(ii) a stair under 1 m wide (Fig. 7.25c) with tapered treads, when used in conjunction with a straight flight, must all be equal in depth (from nosing to nosing) and not less than the goings on the straight flight when measured in the centre of the flight (Fig. 7.25e).

(iii) a stair over 1 m wide (Fig 7.25d) with tapered treads, when used in conjunction with a straight flight, must all be equal in depth (from nosing to nosing), and not be less than the goings on the straight flight when measured 270 mm in from both sides of the strings (see Fig. 7.25g).

Figures 7.25 and 7.26 give two examples for this type of stair incorporating tapered treads.

Figures 7.25a and 7.26a show a stair with a 90° turn (quarter turn) using tapered treads and would be used in a situation where a straight flight would be impractical because of limited space available. Another example is shown in Figs 7.25b and 7.26b, which show a 180° turn (half turn) on the stair. This is ideally suited for a loft conversion as the half turn takes account of the sloping roof and gives sufficient headroom when in use.

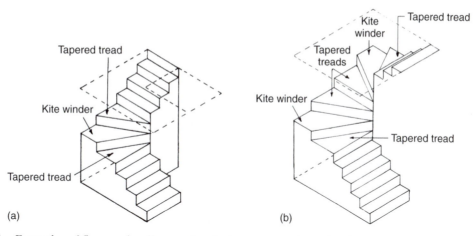

Fig. 7.26 *Examples of floor and well space in relation to tapered tread stairs*

7.3.8 Construction (tapered treads)

Because of possible accidents on this stair type, consideration must be given to Approved Document K in relation to design and fitting.

Furthermore, it is necessary to set out full size all the required true shapes of treads, strings and setting out of the newel posts, to obtain the construction details and required jigs.

Further details can be found in *Carpentry and Joinery: Work Activities* by Brian Porter and Reg Rose (2002), section 11.10.

7.3.9 Shaped ends to bottom treads

If a bottom step protrudes forward of a newel, it will be constructed separately. Figure 7.27 shows how three of these step ends would appear, together with details of how they may be made up, as follows:

(i) *Splayed end (Fig.7.27a):* Riser mitres may be tongued (with a plywood tongue) or screwed together via a ledge. The riser is tongued into and pocket-screwed on to the underside of the tread. The step end is housed into and screwed to the newel.

(ii) *Bull-nosed (quadrant) end (Fig. 7.27b):* The riser is pre-formed either by laminating several layers of wood veneer (plywood) around a former or by bending and gluing thin plywood to a timber backing, with a built-up quadrant block. Fixing of the riser to the tread and of the step to the newel are as above.

(iii) *Half-round end (Fig. 7.27c):* This has a similar construction to a bull-nosed end, except that the step protrudes both forwards and sideways.

In all cases, the underside of the tread will need grooving to receive the riser tongue. This can be done with a portable electric-powered router, as shown in Volume 1, 6.15.

Note: *Once the stair assembly is completed and leaves the workshop it should be protected against any intake of moisture – particularly while in transit, be it to storage or site. Finished surfaces (especially polished hardwoods or softwoods) may need protection against scratches. Softwood rubbing strips or rolls of corrugated card or plastics, held in position with tape or bands, etc., could be used for this purpose.*

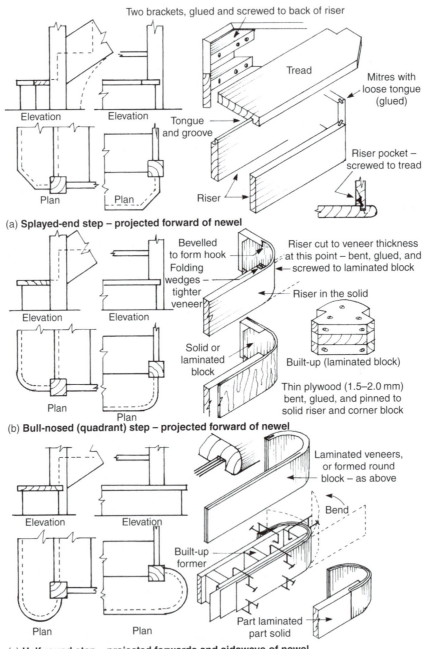

(a) **Splayed-end step – projected forward of newel**

(b) **Bull-nosed (quadrant) step – projected forward of newel**

(c) **Half-round step – projected forwards and sideways of newel**

Fig. 7.27 *Shaped ends to bottom steps*

7.3.10 Alternating treads (Fig. 7.28)

This type of design is only used in situations where there is insufficient space for a conventional stair, for example a loft conversion that requires access to only one habitual room. See Building Regulations Approved Document K.

Each tread has been specially shaped by removing part of the tread (generally a quarter of the width) to form alternating handed treads which must include non-slip surfaces.

The safety aspect of this stair relies on familiarity and regular use. Additionally, a handrail must be included on both sides of the stair.

7.3.10.1 Construction (alternating treads)

The method of construction is similar to conventional stairs where the treads are housed into the strings. However, wedges are generally not included as these would be exposed and would not keep in with the design of the surrounding area.

Treads may be secured by fixing through the strings into the end grain of the treads to prevent spreading by the methods shown in Fig. 7.17.

The open risers must be constructed in such a way that a 100 mm diameter sphere cannot pass through the gap below each tread, even when a 'drop' riser is included, Fig. 7.28 (see also, section 7.3.3 and Fig. 7.18).

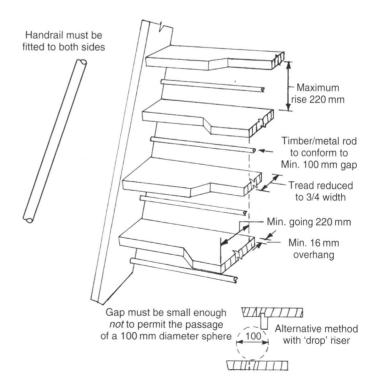

Handrail must be fitted to both sides

Maximum rise 220 mm

Timber/metal rod to conform to Min. 100 mm gap

Tread reduced to 3/4 width

Min. going 220 mm

Min. 16 mm overhang

Gap must be small enough *not* to permit the passage of a 100 mm diameter sphere

Alternative method with 'drop' riser

Fig. 7.28 *Alternating treads*

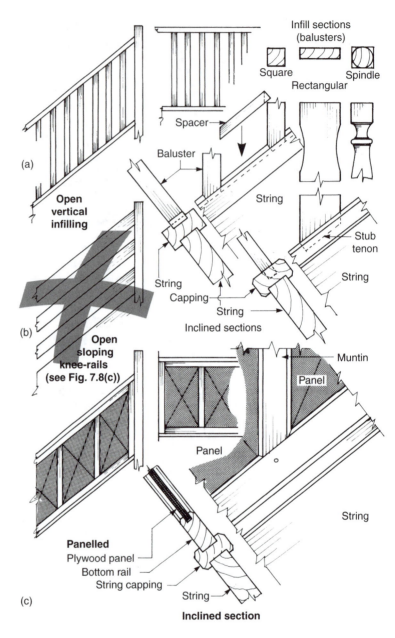

Fig. 7.29 *Types of balustrades*

7.4 **Balustrades**

Balustrades may be described as having either:

(a) an open vertical infilling of balusters;
(b) an open sloping infilling of knee-rails, or
(c) a closed (solid or panelled) infill.

Open vertical balusters (Fig. 7.29a): Square (32 mm × 32 mm), rectangular (32 mm × 61 mm), or spindle (45 mm × 45 mm) sections may be used as balusters. They are mortised or housed with or without 'spacers' into the handrail and the string or string capping.

Table 7.3 Types of balustrade

Type	Advantage	Disadvantage	Remarks
Open vertical balustrades (Fig. 7.29a)	Traditional or modern styles can be employed. Natural and artificial light virtually unrestricted from reaching steps.	Can be expensive. Difficult to clean and maintain, i.e. paint, etc.	Generally safe for children.
Open sloping knee-rails (Fig. 7.29b)	Can be easy to install. Easy to clean and maintain. Only slight restriction in certain circumstances of light to steps.	Style is limited. Mid-rail stiffeners could be used as toe-holds by children.	Unsuitable for landings (see section 7.2.3 and Fig. 7.8c). This type of balustrade (stair and landing) is not to be used in a domestic dwelling.
Panelled (Fig. 7.29c)	Provides privacy. Offers sense of security.	May shut out some natural or artificial light to steps. Can be plain in appearance.	Panels may be used for decorative features, i.e. hardwood, etc.

Open sloping knee-rails (Fig. 7.29b): Three boards of, say, 32 mm × 140 mm section are mortised into or surface fixed on to newels. The knee-rail pattern should not be continued on a landing because of the dangers previously mentioned in section 7.2.3 and Fig. 7.8c).

Panelled (Fig. 7.29c): Usually a series of plywood/MDF panels is fitted into the grooved handrail, string capping, and muntins (vertical members used to divide a large panelled area into smaller panels).

Table 7.3 compares these three types of balustrade.

7.5 Installation (site fixing) (Fig. 7.30)

Items required will include:

- the stair;
- newels (newel posts);
- handrails;
- the bottom step (if of the type which protrudes beyond the newel);
- the top riser and nosing;
- apron linings (linings to a stairwell and bulkhead) (see section 7.6).

The following stages are to be used only as guides – they are not a specific order of installation, as methods and techniques vary according to the stair type, location, site conditions, etc.

(i) Determine the finished floor level (FFL).
(ii) Cut the bottom of the wall string to suit the FFL and the height of skirting board (Fig. 7.30c).

(iii) Cut the top of the wall string to fit onto and against the landing (stairwell) trimmer or trimming joist, with provision to meet the landing skirting board (Fig. 7.30a).

(iv) With the stair temporarily in position, provision can be made for fixing the wall string to the wall by marking its position and plugging the wall. If direct fixing is possible, this stage is left until fitting is completed. The floor position of the lower newel can be marked at this stage.

(v) Fit the pre-shaped and mortised lower newel to the string and make provision for floor anchorage – Fig. 7.31 illustrates several alternatives.

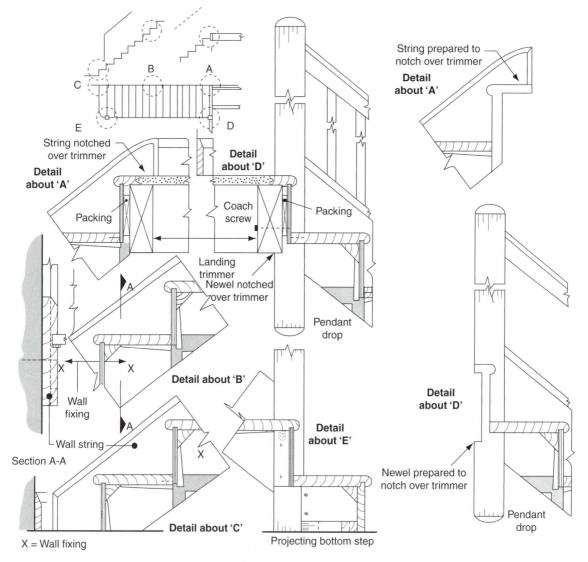

Fig. 7.30 *Construction and fixing details*

(vi) Attach the lower newel to the string using draw-pins (pointed hardwood dowel, see Figs 7.16 and 7.19).

(vii) Fit and fix the pre-made bottom step to the wall string (glue and wedge, Fig. 7.30c), the riser (glue and screw the riser to the tread or the tread to the riser), and the newel (screw the step riser to the newel, Fig. 7.30e).

(viii) Notch the top newel over the trimmer to form a location joint (Fig. 7.30d). If the stairwell is not square with the wall, small allowances may be made at this stage.

(ix) Assemble the top newel to the string and handrail and (if used) the knee-rails to both newels. Joints are glued and drawn together as necessary, draw-board. If balusters are tenoned

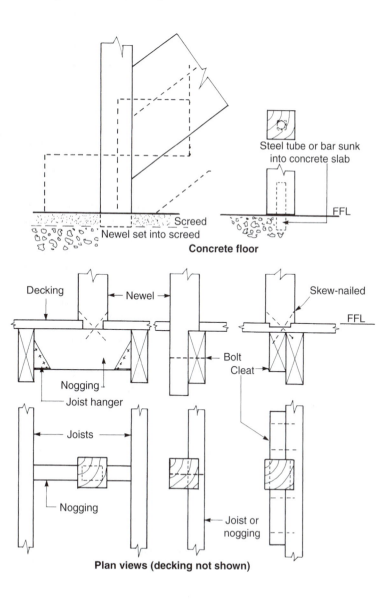

Fig. 7.31 *Bottom newel – post anchorage*

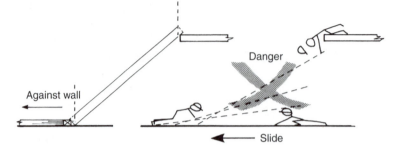

Fig. 7.32   *Providing a temporary stair anchor*

to both the handrail and the string, they too must be included at this stage.

  (x) Fit the top riser and the landing nosing and fix together (glue and pocket-screw, Fig. 7.27a). Glue and screw the assembly to the string, newel housing, and the tread below. Alternatively, the landing nosing and top riser may be fitted after the stair has been fitted in place. The top landing nosing is positioned in place with the top riser pushed up from underneath and 'packed' off the face of the trimmer (Fig. 7.30a). Once fitted, it is then secured to the top tread.

 (xi) Locate the stair into its final position. Check for wall and floor alignment and that the newels are plumb.

(xii) Securely fix the lower newel (the method will depend on the type of floor construction, Fig. 7.31). Fix the top newel from trimmer to newel using a coach screw (see Fig. 7.30(d)) and the landing nosing to the trimmer. Nail or screw (depending on the base material) the wall string to the wall at under tread locations (Fig. 7.30b) – use wood packings at fixing points between the wall and the string if after alignment it is found that there are any gaps; otherwise the stair could be either distorted or damaged.

(xiii) Cut and fix string balustrades with equal gaps, as shown in section 7.8.1.

Landing balustrade assemblies form an attachment with their handrail and floor margin strip (Fig. 7.33a). They may be fixed with the stair or at a later stage – in either case the stairwell opening must not be left without guard rails.

Note: *Before the stair is tried (reared up) in its intended position, a securely anchored ground stay (kicker) will be required at the foot of the stair as a precaution against the stair sliding, as shown in Fig. 7.32*

7.6 Apron linings and bulkheads

The exposed edges of the stairwell can be lined with either solid timber or plywood (Fig. 7.33a). The lining (apron) is packed off the trimming to provide edge support to a margin strip that edges the landing decking and may provide a fixing for balusters. The under edge, which could act as an edge for the plasterer, is covered by a wood trim.

In situations where the room above a stair extends forward of the stairwell, a bulkhead similar to that shown in Fig. 7.33b may be formed.

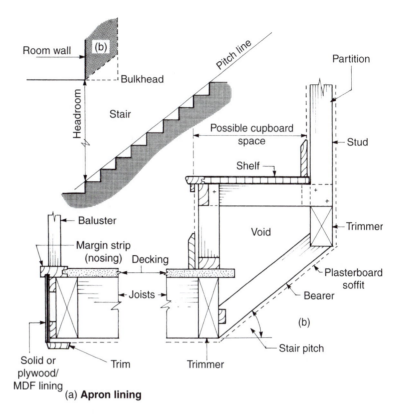

Fig. 7.33 *Stairwell lining and bulkhead construction*

7.7 Protection

Once installation is complete, treads (particularly their nosings) can in some cases be protected from the traffic of operative's footwear by tacking thin strips of wood, plywood, or hardboard (rough side up) to the surfaces.

Newels can easily be damaged by the movement of building materials (lengths of timber, pipes, etc.). This risk can be reduced by wrapping them in card or plastics and/or tying narrow strips of hardboard or plywood, etc. around them with tape or bands.

Handrails may need similar protection.

7.8 Stair calculations

Before any templates (pitch board, etc.) or jigs (a cramp-on guide for repetitive work, i.e. cutting housings, Figs 7.10 and 7.14) can be made, the rise and going for each step must be calculated from the information obtained from the site visit (see section 7.2.1).

A stair may be designed to be constructed to use the minimal amount of materials and space, however at the same time it must conform to current Building Regulations (Approved Document K).

A situation may arise where the subdivisions (number of rises) of the total rise and going may not be exactly equal to the total rise or going of the original measurements from site, in which case they are adjusted accordingly.

The final measurement may be 1 to 5 mm over or under – this difference in measurement can be, in some cases, lost in the fitting and the cleaning-up process. However, amounts 5 mm or over may prove to be more difficult.

Example calculation for a domestic (private) dwelling

Let's assume a total rise of 2.52 m from FFL to FFL (finished floor level) and a total going from the face of the first riser to the face of the last riser of 3.25 m. These are the measurements obtained from site.

Using a calculator (similar to Fig. 7.35) and the maximum step rise of 220 mm (as shown in Tables 7.1 and 7.4), divide this figure into the total rise of 2520 mm; this will give you 11.454. We can round this up to 12.

We can now divide the rise (2520 mm) by 12 to give 210 mm, which is below the maximum of 220 mm. Because the stair in question now has 12 risers, it will require 11 goings (always one less than the number of risers).

Therefore, to summarize, a single riser height can be calculated as follows:

$$2520 \div 220 = 11.454 \text{ (round up to 12 risers)}$$

$$2520 \div 12 = 210 \text{ mm (for each riser)}$$

So this particular stair has 12 risers at 210 mm with 11 goings (whose size is yet to be determined). The maximum angle for a private stair is 42° (see Table 7.1). To be on the safe side an angle of 41.9° would be preferable to use.

At this point you will require a scientific calculator – alternatively refer to section 7.3.1, Setting-out.

The tangent of an angle (Tan) (Fig. 7.34)

This is the result of the ratio of the opposite and adjacent sides in a right-angled triangle. If the angle and the length of one of the sides are known it is possible, by the use of trigonometry, to obtain the unknown length of the other side.

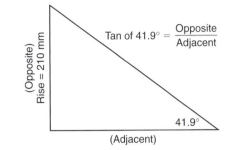

Fig. 7.34 *The tangent of an angle*

Alternatively, if the length of the opposite and adjacent sides are known, the angle can be determined by rearranging (transposing) the formula (Fig. 7.34).

To find the going (adjacent), the formula for the tangent (Tan) of an angle can be used but will need transposing (rearranging) a number of times until the required formula to find the adjacent is obtained.

As the rise and angle is known, the formula of Tan, which is the result of the length of the opposite side of the known angle divided by the length of the adjacent side of the known angle, can be used.

This is written first as:

$$\text{Tan} = \frac{\text{Opposite}}{\text{Adjacent}}$$

First, as the adjacent is the unknown factor, we arrange the formula so that both sides are required to be multiplied by the adjacent.

Therefore:

$$\text{Tan} \times \text{Adjacent} = \frac{\text{Opposite}}{\text{Adjacent}} \times \text{Adjacent}$$

Because the right-hand side has two adjacents, they can both be cancelled out.

Therefore:

$$\text{Tan} \times \text{Adjacent} = \frac{\text{Opposite}}{\cancel{\text{Adjacent}}} \times \cancel{\text{Adjacent}}$$

Which leaves us with:

$$\text{Tan} \times \text{Adjacent} = \text{Opposite}$$

The left-hand side now contains Tan, which we need to remove as we only want the adjacent.

Note: *Remember that any series of adjustments must be kept in balance, i.e. both sides!*

We can do this by dividing both sides by Tan.

Tan is placed under the formula (because we are dividing) on both sides in order to keep in balance.

Therefore:

$$\frac{\text{Tan} \times \text{Adjacent}}{\text{Tan}} = \frac{\text{Opposite}}{\text{Tan}}$$

Now as Tan appears twice on the left-hand side, they cancel each other out.

Therefore:

$$\frac{\cancel{\text{Tan}} \times \text{Adjacent}}{\cancel{\text{Tan}}} = \frac{\text{Opposite}}{\text{Tan}}$$

So finally, we are left with our required formula for determining the adjacent.

This now reads:

$$\text{Adjacent (going)} = \frac{\text{Opposite}}{\text{Tan}}$$

Therefore:

$$\text{Adjacent (going)} = \frac{210}{\text{Tan } 41.9°}$$

Note: *Ensure that the calculator is set to degrees (DEG)*

Using a scientific calculator (Fig. 7.35), the Tan of 41.9° is required.

This is carried out by typing 41.9 and then pressing the Tan button (Fig. 7.35) = 0.897248684.

Therefore to find the going, the rise of 210 is divided by 0.897248684 = 234.048 mm, or 235 mm (rounded up)

To summarize,

$$\text{Going} = \frac{210}{.897248684} = 235 \text{ mm}$$

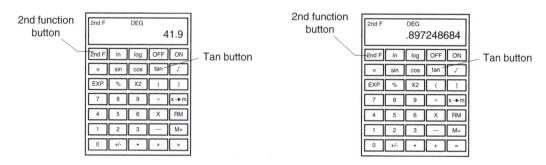

Fig. 7.35   *Using a calculator to obtain the TAN of 41.9°*

Table 7.4 Combinations of rise and going for private stair

Rise (mm)	*With*	Going (mm)
(a) 155 to 220		245 to 260
	or	
(b) 165 to 200		223 to 300

Finally, with this calculation, the stair has:

12 risers at 210 mm, with 11 goings at 235 mm, with a pitch of 41.9°

However, we must now check the aggregate (the amount or sum of the ratio 2 × R + G must be between 550 and 700).

Their aggregate, according to the rule would therefore be:

$$2 \times 210 + 235 = 655 \, mm$$

As 655 mm is not less than 550 mm nor more than 700 mm, this will be acceptable provided that the angle formed by the going and the pitch line (hypotenuse) is between the acceptable limits of 32° and 42° – which it is.

Finally, we can carry out the following checks:

To check the risers against the total rise measurement,

$$210 \times 12 = 2520 = \text{original total rise.}$$

To check the goings against the total going measurement,

$$235 \times 11 = 2585 = \text{below original total going of 3250,}$$
$$\text{which is satisfactory.}$$

The angle 41.9° used in this calculation is under the maximum acceptable limit of 42° for combinations of rise and going within the following ranges shown at (a) in Table 7.4.

Example stair calculation under the category for institutional and assembly

For a stair that comes under Institutional and Assembly (where people gather), for example a hospital or hotel, the calculations are similar to a private stair, but by using the maximum riser height and minimum going (Table 7.5) the angle achieved will always be just over 32° as opposed to a private stair (Table 7.4b), where there is a possibility of exceeding the maximum angle of 42°.

It should not, however, be less than 30° as this could prove to be too tiring when walking up the stair; for example, by using the minimum rise of 135 and maximum going of 340 it will give an angle of 21° which is far too low.

Assuming a total rise of 2.555 m with a total going of 4.5 m

Table 7.5 Combinations of rise and going for Institutional & Assembly stair

Rise (mm)	*with*	Going (mm)
135 to 180		280 to 340

Therefore:

> 2555 divided by the maximum rise of 180 = 14.194
>
> = 15 (rounded up)
>
> 2555 divided by 15 = 170.3333 (15 risers at 170.3333 mm)

0.3333 of a millimetre is so small that it would be difficult to include in the measurements so the rise could be rounded up to 171.

To check to see how it compares to the original total rise of 2555,

$$171 \times 15 = 2565$$

This now means we are *oversize* by 10 mm, which would be difficult to lose during fitting.

Therefore:

Let's try 170 mm, but by checking the measurements again,

$$170 \times 15 = 2550$$

We are now *under* by 5 mm.

Another way would be to use

$$170.5 \times 15 = 2557.5$$

That now means we are over by only 2.5 mm, even better!

Again, this may be adjusted on site after fixing.

To summarize:

> 2555 ÷ 180 (max rise) = 14.194 (round up to 15)
>
> 2555 ÷ 15 = 170.3333 (round up to 171)
>
> 15 × 171 = 2560 (oversize by 10 mm)

Try:

> 15 × 170 = 2550 (undersize by 5 mm, but may be lost when fixing on site)

Try adding 0.5 mm to 170 mm

15 × 170.5 = 2557.5

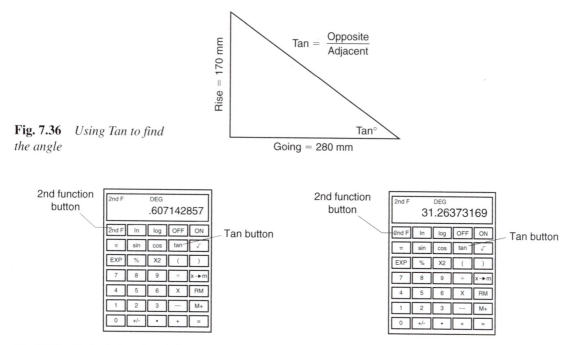

Fig. 7.36 *Using Tan to find the angle*

Fig. 7.37 *Calculating the angle*

We are now oversize by 2.5 mm, which is easily lost when fixing on site. However, 0.5 mm is difficult to work with so in this example we will use 170 mm for each riser. This will mean we are under by 5 mm on the total rise (2.555 m). But as stated previously, this may be adjusted when fitting the stair on site.

To check the angle the required minimum going of 280 mm (Table 7.5) is divided into 170 (rise) using the scientific calculator (Fig. 7.37) and the formula for Tan (Fig. 7.36).

170 divided by 280 = 0.607142857. Now by pressing the 2nd function on the calculator, then the Tan button (Fig. 7.37), the angle obtained is just over 31°.

To summarize, the stair has 15 risers at 170 mm with 14 goings at 280 mm. According to the rule of twice the rise plus the going (2 × R + G) should be between 550 and 700, their aggregate would be:

$$2 \times 170 + 280 = 620 \, mm.$$

As 620 mm is not less than 550 mm nor more than 700 mm, this will be acceptable provided that the angle formed by the going and the pitch line (hypotenuse) is between the acceptable limits of around 30° to 32°, which it is.

7.8.1 Calculating the number of balustrades and the spaces in-between (Fig. 7.38)

Once the stair has been fitted and the handrail is in place, the balusters are cut and fixed into the underside of the handrail (Fig. 7.20 and 7.29), and into the string or string capping (or in the case of landings, the landing capping).

In this example we have 12 goings at 258 mm with 95 × 95 mm newel posts and we will use 32 × 32 mm spindles. The horizontal gap measured between each of the vertical balusters is obtained first by adding together the total step goings measured between the newels of the stair (Fig. 7.38).

As the top and bottom riser faces are generally central on the newel post, a deduction of 2 × ½ the thickness of the newel post is taken off the total goings.

For example, there are 12 goings at 258 = 3096 − 2 × ½ the thickness of newel, i.e. 95. This will leave 3001 mm as shown in Fig. 7.38 (spindle run).

As the Building Regulations stipulate that a 100 mm sphere cannot pass through any openings, this needs to be included when calculating the final gap size.

To obtain the number of balusters, the formula used would be:

$$\frac{\text{Spindle run}}{\text{Spindle width} + 100}$$

Example (Fig. 7.38)

The diagram in Fig. 7.38 shows two examples where spindles are to be fitted and spaced equally, either on the stair or along the landing. In both cases, the horizontal run must be obtained first.

Therefore:

Spindle run = 3001 mm, spindle width = 32 mm.

$$\frac{3001}{32 + 100} = \frac{3001}{132} = 22.7348$$

The figure is now rounded up to give *23 spindles with 24 gaps* (always one more gap than the number of spindles).

To obtain the exact spacing between each baluster, the spindle widths are multiplied by the number of spindles required. This figure is then deducted from the spindle run.

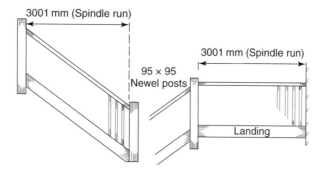

Fig. 7.38 *Calculating spindle gaps*

Finally, the result is divided by the number of gaps.

Therefore:

$$\frac{\text{Spindle run} - (\text{spindle widths} \times \text{no. of spindles})}{\text{no. of gaps}}$$

Example

Using the information above:

$$\frac{3001 - (32 \times 23)}{24} = \frac{3001 - 736}{24} = 94.375$$

The result is 23 spindles with 24 gaps spaced at 94.3 mm.

Finally, to recap, stair calculations are not always straightforward and do require figures to be adjusted to suit the situation. A copy of the current Building Regulations for consultation is always required when undergoing this task.

A stair usually includes a capping piece (Fig. 7.29), which is wider than the thickness of the string by around 12 mm, and grooved (usually 6 to 8 mm deep) both sides along its length. The groove underneath the capping is designed to fit over the top edge of the string, and once cut is secured in place, whereas the groove along the top edge (whose width is the same thickness as the balusters) will allow the spacing and fixing of the balustrade. Usually a chamfer or pencil round is included along the top edges of the capping.

7.8.2 Calculating spacer length between balusters (spindles) using a stair of 41.9°

Once the balusters are fixed in place, it is usual practice to fill in the exposed groove between each of the balusters with a spacer.

For this exercise we are using the horizontal space between the balusters obtained in section 7.8.1 above. To obtain the true length of the

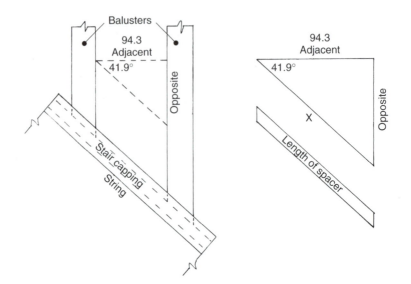

Fig. 7.39 *Calculating spacer length between balusters*

spacers, and assuming the gaps are equal, calculations using Tan and Pythagoras theorem are used together with the horizontal measurement between each baluster (see section 7.8.1) and the angle of the stair (Fig. 7.39).

Before we can obtain the length of X (the true length of the spacer) the length of the opposite side needs to be found first.

To find the length of the opposite side, the formula for the tangent (Tan) of an angle can be used, but will need transposing (rearranging) until the required formula to find the opposite is obtained.

As the adjacent side and angle is known, the formula of Tan, which is the result of the length of the opposite side of the known angle divided by the length of the adjacent side of the known angle, can be used.

Note: *Remember any series of adjustments, must be kept in balance, i.e. both sides!*

This is written first as:

$$\text{Tan} = \frac{\text{Opposite}}{\text{Adjacent}}$$

First, as the opposite is the unknown factor, we arrange the formula so that both sides are required to be multiplied by the adjacent.

This is written here as:

$$\text{Tan} \times \text{Adjacent} = \frac{\text{Opposite}}{\text{Adjacent}} \times \text{Adjacent}$$

Because the right-hand side has two adjacents, they can both be cancelled out.

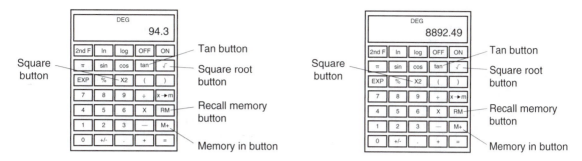

Fig. 7.40 *Finding the square of a given figure*

Therefore:

$$\text{Tan} \times \text{Adjacent} = \frac{\text{Opposite}}{\cancel{\text{Adjacent}}} \times \cancel{\text{Adjacent}}$$

This leaves us with:

$$\text{Tan} \times \text{Adjacent} = \text{Opposite}$$

We now have the required formula to obtain the length of the opposite side

Therefore:

$$\text{Tan } 41.9° \times 94.3 = \text{Opposite side}$$

Using the calculator we first find the Tan of 41.9° as before in Fig. 7.35, which is 0.897248684, by inserting 41.9 and pressing the Tan button.

Now we multiply 0.897248684 by 94.3 which gives us 84.6 – the length of the opposite side.

Using Pythagoras' theorem below, we can obtain the length of X – the length of the spacer (Fig. 7.39).

'The square on the hypotenuse of a right-angled triangle (X) is equal to the sum of the squares of the opposite two sides.'

The term 'square' refers to the length of each of the sides multiplied by themselves.

For example $94.3 \times 94.3 + 84.6 \times 84.6$.

The calculator can be used for this operation as it contains a 'square' button (Fig. 7.40).

First type in 94.3 and then press the X^2 button to obtain 8892.49.

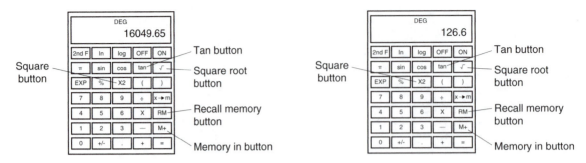

Fig. 7.41   *Obtaining the true length of the spacer*

This figure is stored in the calculator by pressing the M+ button. Repeat the same process for 84.6.

Both results are now stored.

To recall the results, the recall memory (RM) button is pressed to give 16049.65 (Fig. 7.41).

We now need to find what number multiplied by itself will give us 16049.65.

This is known as the '*square root*' of a number (Fig. 7.41).

The square root button is finally pressed to give us the true length of the spacer, which is 126.6 mm.

References

Building Regulations, Approved Document K, Protection from falling, collision and impact, section K1: 2000.

Building Regulations, Approved Document A, Structure: 2000.

Building Regulations, Approved Document D, Toxic substances: 2000.

BS 5395: Part 1: 2000, Stairs, ladders and walkways. Code of practice for the design, construction and maintenance of straight stairs and winders.

BS 7956: 2000, Specifications for primers for woodwork.

BS 4071:1996 (amended 1992), Specification of polyvinyl acetate (pva) emulsion for wood.

BS EN 923, 2005 Adhesives. Terms and definitions.

BS EN 12436: 2002, Adhesives for load-bearing timber structures. Casein adhesives. Classification and performance requirements.

BS EN 1186–2, 1988, Timber for, and workmanship in joinery. Specification for workmanship.

BS EN ISO 4618–3:2000, Paints and varnishes. Terms and definitions for coating materials. Surface preparation and methods of application.

HSE (Health & Safety Executive) Manual handling leaflets.

8

Wood wall trims and finishes

With the exception of the 'Delft' shelf, which is sometimes called a 'Plate' shelf, examples as to where we might find wood trims and finishes are shown in Fig. 8.1. Table 8.1 can be used to identify and locate the most common of these wall trims and finishes so as to ascertain whether they have a functional or decorative role, or possibly both.

Take note that architraves and skirting boards have several functions; these include:

(a) Concealing gaps which are likely to occur as a result of moisture or thermal movement from abutting dissimilar materials;

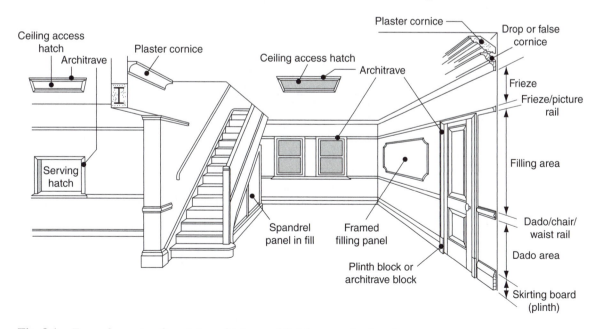

Fig. 8.1   *Examples as to where internal trims and finishes may be found*

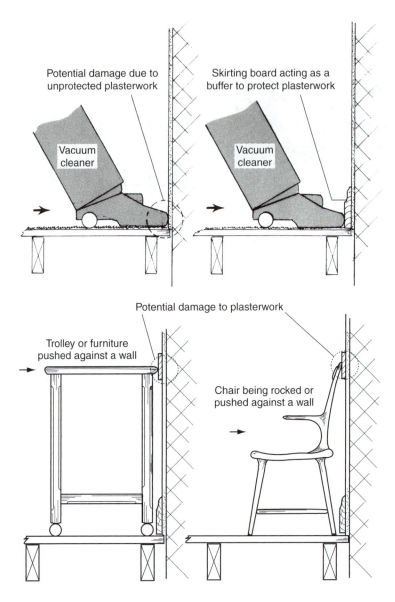

Fig. 8.2 *Wall trims used to protect plasterwork*

(b) Protecting plasterwork from the occasional knock at floor level by floor cleaning machines;

(c) Providing a decorative finish, and in some cases a decorative feature.

Dado rails on the other hand can:

(a) Physically protect plasterwork at or about waist height when moving furniture or chairs up against a wall;

(b) Serve solely as a decorative feature.

Where trims are used, or serve as a means to protect a wall's surface (plasterwork, see Fig. 8.2), wood or wood composites (section 8.1.1)

Table 8.1 Wall trims and finishes (Mouldings) in relation to their use

Component	Functional	Decorative	Location
Architrave	Yes (P)	Yes	Wall to door/hatch lining Ceiling to hatch lining
Skirting board	Yes (P)	Yes	Wall to floor
Plinth blocks (Fig. 8.17)	Yes (P)	Yes	Joints between architrave and skirting board
Corner block (Fig. 8.13)	Yes (P)	Yes	Corner joints between architrave legs & head
Dado (chair or waist) rail (Fig. 8.2)	Yes/No	Yes (P)	Wall trim (830–900 mm) above floor level
Picture (frieze) rail (Figs 8.1, 8.5 and 8.31)	Yes/No	Yes (P)	Wall trim (400–500) below ceiling level
Delft (plate) shelf (Fig. 8.35)	Yes (P)	Yes	Wall trim (400–500) below ceiling level
Cornice (drop or false) (Fig. 8.37)	Yes (P)	Yes	Wall trim – can be used to house concealed lighting units
Infill panel frame (Fig. 8.39)	No	Yes (P)	Wall trim – mid-way between picture rail & dado rail

'P' = primary use.

are the ideal medium for this purpose because they are much more resilient to knocks than plasterwork.

Decorative properties: with the exception of a specific 'chair rail' (Fig. 8.2), all the other items listed in Table 8.1 will be shaped as a moulding to be in keeping with the design and style of the room.

8.1 Mouldings (shaped sections)

Unlike shaped sections used for doors and windows (section 4.5.3, Fig. 4.15 and section 3.2.1, Fig. 3.6) the shaped edge or edges of moulded trims do not always reflect their end use. Some do, however, have recognizable profiles, which carry a common name recovered from a bygone age of the classical periods. It is these classical lines (see Volume 1, section 1.9.4) that we tend to copy when reproducing the various shapes we call mouldings.

Figure 8.3 shows how classical shapes based on the Roman and Grecian styles can be reproduced using simple geometrical means. Roman styles are based on the arcs of a circle (see Volume 2, section 4.1), whereas Grecian styles may be based on the softer curves of the ellipse, parabola and hyperbola formed as a result of conic sections. Figure 8.4 shows how conic section are derived by taking vertical and sloping sections though a cone – hence the term 'conic' sections.

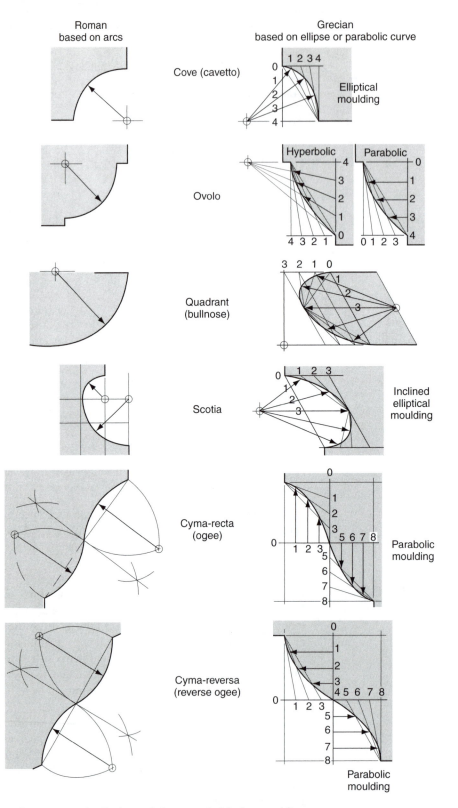

Fig. 8.3 *Drawings geometrically formed shapes suitable for mouldings*

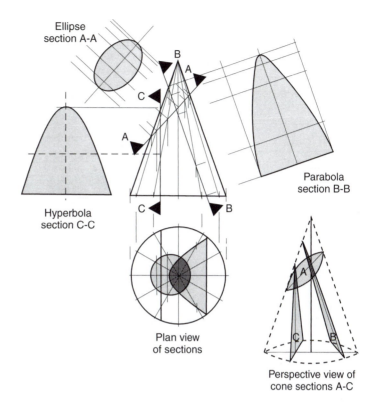

Ellipse
section A-A

Parabola
section B-B

Hyperbola
section C-C

Plan view
of sections

Perspective view of
cone sections A-C

Fig. 8.4 *Shapes derived from sections of a cone (conic sections) used to form Grecian mouldings*

Commercially produced shaped sections (profiles) for architraves and skirting boards are generally limited to a small number or design profiles; similarly repeated sections are however available in various sizes. Examples of these are shown in Fig. 8.5, together with profiles for dodo and picture rails and various panel mouldings.

8.1.1 Materials used for mouldings

If a painted finish is required it can be most cost effective to use clear grades of softwood, preferably Redwood (dead and large live knots must be avoided on the moulded portion). Medium density fibreboard (MDF) (Vol. 1, section 4.6.6 change) mouldings are a more modern and popular alternative, providing an unblemished surface usually pre-painted as part of the manufacturing process. MDF profiles are also available covered in a variety of different films, to simulate decorative hardwoods. The use of hardwood moulded profiles is usually reserved for small decorative features and finishes.

8.1.2 Forming the shaped profile

Propriety, or commercially produced shaped sections used as mouldings are produced either by:

(a) Moulding machine;
(b) Spindle moulder (vertical spindle or shaping machine);

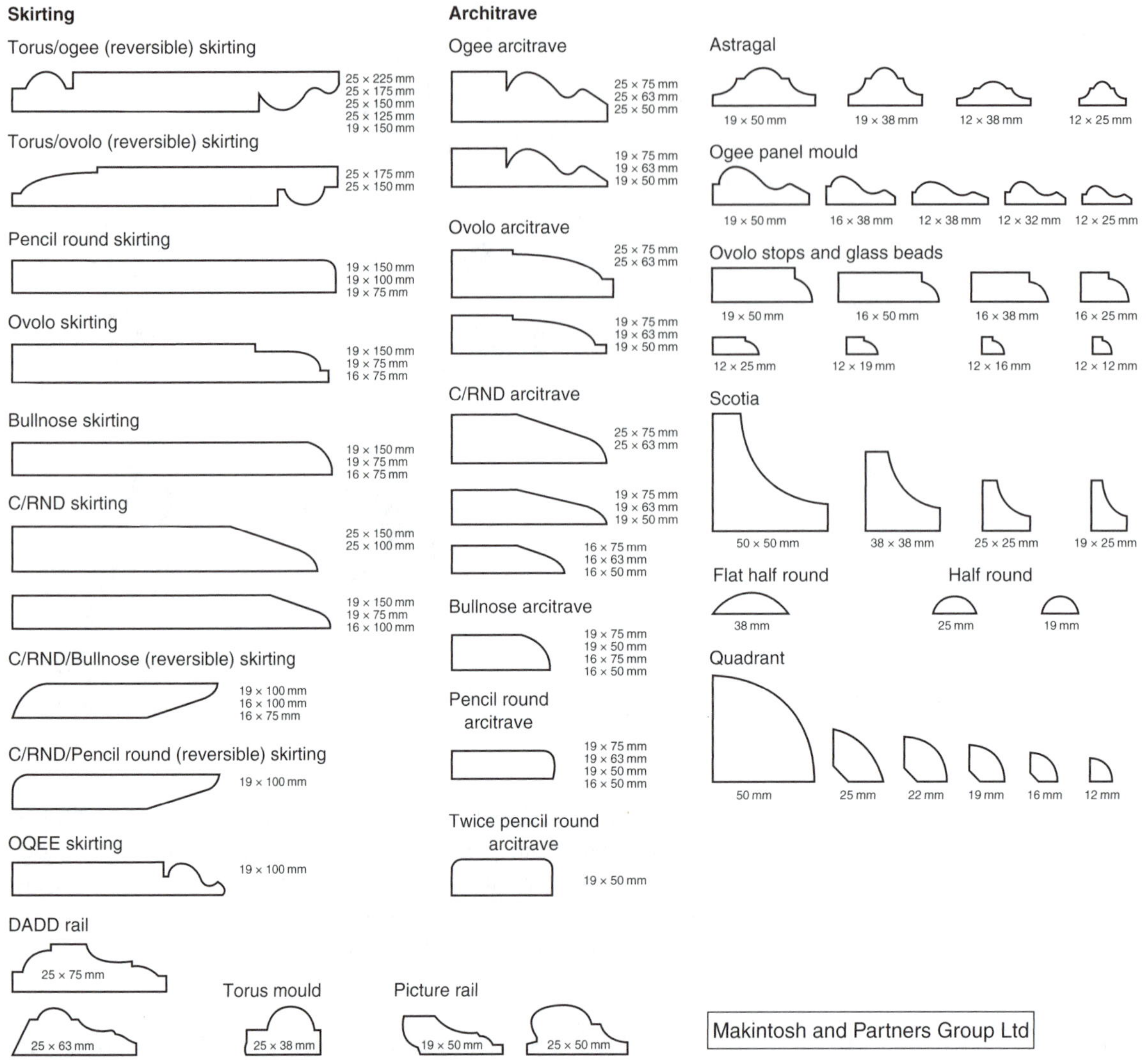

Skirting

Torus/ogee (reversible) skirting

25 × 225 mm
25 × 175 mm
25 × 150 mm
25 × 125 mm
19 × 150 mm

Torus/ovolo (reversible) skirting

25 × 175 mm
25 × 150 mm

Pencil round skirting

19 × 150 mm
19 × 100 mm
19 × 75 mm

Ovolo skirting

19 × 150 mm
19 × 75 mm
16 × 75 mm

Bullnose skirting

19 × 150 mm
19 × 75 mm
16 × 75 mm

C/RND skirting

25 × 150 mm
25 × 100 mm

19 × 150 mm
19 × 75 mm
16 × 100 mm

C/RND/Bullnose (reversible) skirting

19 × 100 mm
16 × 100 mm
16 × 75 mm

C/RND/Pencil round (reversible) skirting

19 × 100 mm

OQEE skirting

19 × 100 mm

DADD rail

25 × 75 mm

25 × 63 mm

Torus mould

25 × 38 mm

Architrave

Ogee arcitrave

25 × 75 mm
25 × 63 mm
25 × 50 mm

19 × 75 mm
19 × 63 mm
19 × 50 mm

Ovolo arcitrave

25 × 75 mm
25 × 63 mm

19 × 75 mm
19 × 63 mm
19 × 50 mm

C/RND arcitrave

25 × 75 mm
25 × 63 mm

19 × 75 mm
19 × 63 mm
19 × 50 mm

16 × 75 mm
16 × 63 mm
16 × 50 mm

Bullnose arcitrave

19 × 75 mm
19 × 50 mm
16 × 75 mm
16 × 50 mm

Pencil round arcitrave

19 × 75 mm
19 × 63 mm
19 × 50 mm
16 × 50 mm

Twice pencil round arcitrave

19 × 50 mm

Picture rail

19 × 50 mm   25 × 50 mm

Astragal

19 × 50 mm   19 × 38 mm   12 × 38 mm   12 × 25 mm

Ogee panel mould

19 × 50 mm   16 × 38 mm   12 × 38 mm   12 × 32 mm   12 × 25 mm

Ovolo stops and glass beads

19 × 50 mm   16 × 50 mm   16 × 38 mm   16 × 25 mm

12 × 25 mm   12 × 19 mm   12 × 16 mm   12 × 12 mm

Scotia

50 × 50 mm   38 × 38 mm   25 × 25 mm   19 × 25 mm

Flat half round   Half round

38 mm   25 mm   19 mm

Quadrant

50 mm   25 mm   22 mm   19 mm   16 mm   12 mm

Makintosh and Partners Group Ltd

Fig. 8.5 *Commercially available shaped sections (profiles) – mouldings*

(c) Embossing (pressing);
(d) Machine carving (using computer-aided-design 'CAD').

Purpose-made mouldings are produced either by:

(a) Spindle moulding (this machine is outside the scope of this series of books. Special training and certification is required before operating a spindle moulding machine);
(b) Portable powered hand router with shaped cutters (Volume 1, section 6.15);
(c) Hand tools (including moulding planes) (Fig 8.7);
(d) Combination and multi-planes (universal plane) (Fig 8.8).

8.1.3 Uniformity of commercially produced mouldings

If you purchase pre-formed mouldings from a supplier you expect uniformity throughout that particular pattern. However, the finished shaped section can vary between each run-off or batch, and between different suppliers, so don't be surprised if the profile does not match your previous order. These variations are usually due to slight differences in cutter profile. It is always advisable to use the same supplier for mouldings and order enough from the same run-off or batch.

Note: *Even though architraves and skirting boards differ in sectional size it is common practice to use the same moulded profile in the same room*

8.1.4 Purpose-made mouldings

Almost any moulding can be produced or reproduced with a portable powered router (Volume 1, section 6.15), providing the appropriate cutters are available. For large profiles the router will be hand held and passed over the work piece via the appropriate fence or jig. For smaller moulding profiles of a short run-off, the router may be inverted into a purpose-made table and used as a small spindle moulder (Volume 1, section 6.15.4).

Working mouldings using hand tools can be a slow process, but there are occasions that warrant this – usually when a short length, non-standard moulding is required due to damage, or when it is impracticable to start making up a machine cutter to suit. Hand working means using all the tools at your disposal; this may include using portable powered hand tools such as the router and planer – not only can they speed up the process but they can, used correctly, add to detailing accuracy. Figure 8.6 shows how a simple parabolic ovolo moulding profile, often used on the edge of skirting boards, can be reproduced using hand tools, as follows:

(a) Ensure the sample is clear of paint build-up (Fig 8.6a) – remove as necessary.
(b) Make a template of the profile out of card (Fig 8.6b), or use a proprietary needle template former (Fig. 8.6c).
(c) Mark off the end profile of the section to be copied (Fig 8.6d).
(d) Score gauge lines along all the square edges to be formed on the face side and face edge (Fig. 8.6e) – this will help when setting width gauges of the moulding tools.
(e) With the stock (base material, for example a length of timber) held vertically in the vice and using a plough plane or portable powered hand router cut a groove to form the quirk (Fig. 8.6f).
(f) With the stock held securely flat, form as many grooves as practicable to the contour depth of the prescribed moulding profile (Fig. 8.6g).
(g) Gradually remove remaining waste with bench planes (Fig. 8.6h) whilst constantly checking for the finished profile with the template (Fig. 8.6b).
(h) Using step down grades of abrasive papers and shaped sanding blocks (Fig. 8.6i) remove all the flats left by the planing.

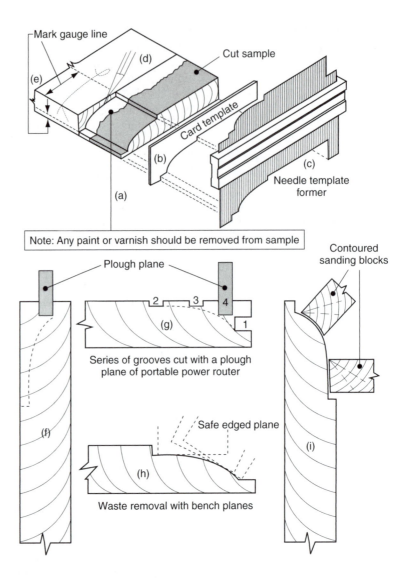

Fig. 8.6 *Copying an ovolo section using hand tools*

This method can be adapted to suit many different profiles, but there will be occasions when inward curves (coves) will need to be formed. In these cases special wooden moulding planes are used (Fig. 8.7).

There are instances when a complicated or intricate piece of moulding would be difficult to accurately produce by moulding planes or even an electric router. In which case, as shown in Fig. 8.8, the 'Clifton' multi-plane together with its range of cutters can for short run-offs produce very good results, and in many situation be easier to set up than the router.

Scratch stock: two examples of these purpose-made devices are shown in Fig. 8.9. They are ideal for cutting (scratching) small moulding

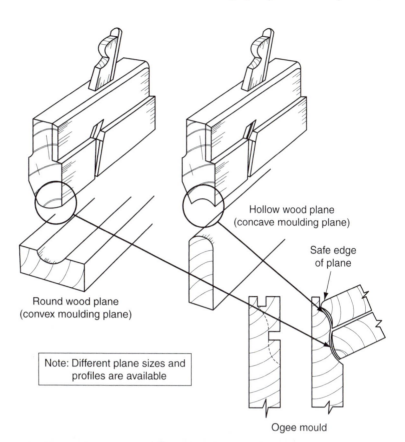

Fig. 8.7 *The purpose and application of traditional wood moulding planes*

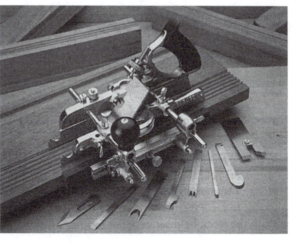

Fig. 8.8 *'Clifton' universal multi-plane*

details, particularly when applied to dense hardwoods. Figure 8.9a follows the traditional pattern, whereas Fig. 8.9b shows one of my modified designs.

The blade can be made from an old saw blade, which is shaped by grinding or filing with a cutting (scraping) angle of about 80° to suit

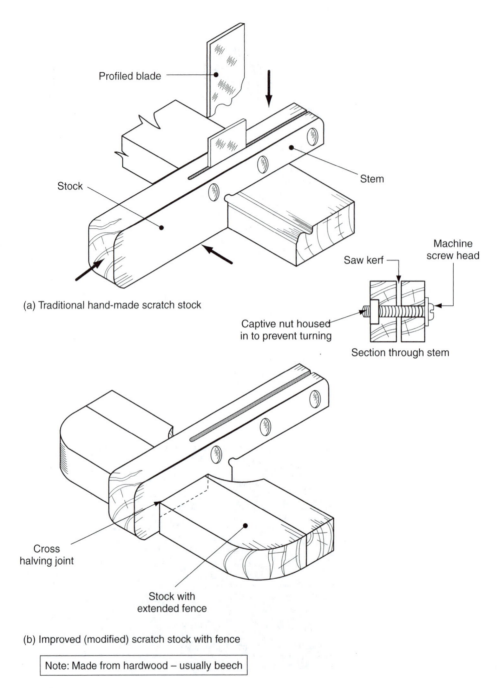

(a) Traditional hand-made scratch stock

Profiled blade

Stock

Stem

Saw kerf

Machine screw head

Captive nut housed in to prevent turning

Section through stem

Cross halving joint

Stock with extended fence

(b) Improved (modified) scratch stock with fence

Note: Made from hardwood – usually beech

Fig. 8.9   *The purpose-made 'scratch stock'*

a required profile. In use the scratch stock (fence) must be held square against the edge of the timber whilst pushing the cutter forward and at the same time lowering it very slowly into the wood – it will take a little practise to perfect this technique. First cuts should always be made into off-cuts of the same wood species.

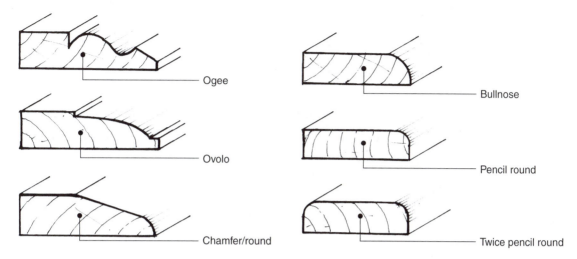

Fig. 8.10 *Commonly available moulded sections used as architraves*

8.2 Architraves

Architraves are more usually associated with forming a frame around internal door linings, but in fact architraves are also used to frame wall and ceiling hatches and some window linings. In all cases they cover the joint between the plasterwork and the wood lining or reveal. Some common profiles are shown in Fig. 8.10, also see Fig. 8.5.

As shown in Fig. 8.11 several methods of framing an opening with architrave can be used to frame an internal door lining, but before describing the different fixing methods we should consider which joint to use and why.

As shown in Fig. 8.11a and 8.11b, the simplest method of joining the head to the legs is by using a mitre joint. The only problem here is you cannot conceal any gaps, which more often than not will (as shown in Fig. 8.12) occur as the timber loses yet more moisture – chances are you will have noticed the effects of this in your own home without realizing what caused it. Methods of reducing the effect of shrinkage include using *corner blocks* at the head, and *plinth blocks* at floor level where they abut the skirting board.

Corner blocks: another way of joining the head to the legs is by using a corner block as shown in Fig. 8.11c, 8.11d, 8.12 and 8.13 featuring a Yorkshire Rose to show how these may be personalized but, as shown in Fig. 8.12, care should be taken to avoid any potential block shrinkage by either housing the architrave into the block or using a material of minimal shrinkage such as MDF. Figure 8.10e shows the use of an ornate pediment head; again, it would be advisable for this to house the tops of the legs to conceal any shrinkage.

Plinth blocks: as shown in Fig. 8.1, 8.11b, d and e when the thickness of the architrave is greater than the thickness of the skirting board,

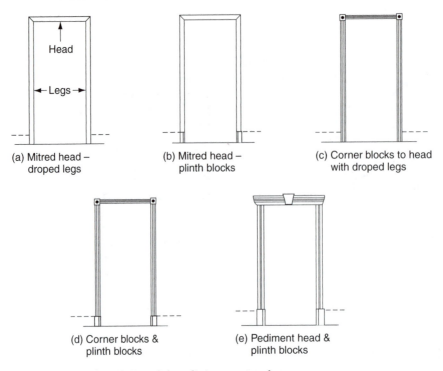

(a) Mitred head – droped legs

(b) Mitred head – plinth blocks

(c) Corner blocks to head with droped legs

(d) Corner blocks & plinth blocks

(e) Pediment head & plinth blocks

Fig. 8.11 *Methods of framing an internal door lining or reveal*

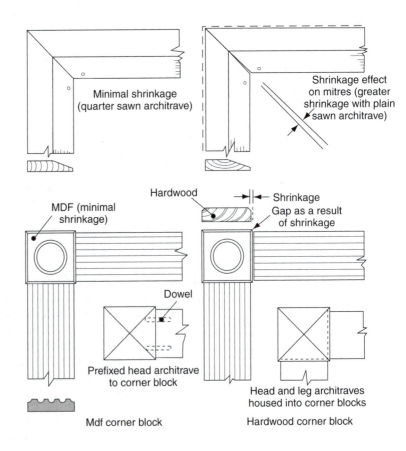

Minimal shrinkage (quarter sawn architrave)

Shrinkage effect on mitres (greater shrinkage with plain sawn architrave)

MDF (minimal shrinkage)

Hardwood

Shrinkage Gap as a result of shrinkage

Dowel

Prefixed head architrave to corner block

Head and leg architraves housed into corner blocks

Mdf corner block

Hardwood corner block

Fig. 8.12 *Shrinkage at corners*

legs can extend to the finished floor level and butted against it. If, on the other hand, the architrave is thinner than the skirting board or is of an ornate nature liable to become damaged at about floor level, it is advisable to incorporate a solid block (although usually shaped with a chamfer to follow the outline of the architrave allowing the door to open beyond 90°) of timber or MDF known as a plinth or architrave block (three examples are shown in Fig. 8.11b, 8.11d, and 8.11e; also see section 8.2.4, Fixing plinth blocks).

8.2.1 Ordering architrave

The amount (linear metres) required for each door opening will depend on:

(a) The opening size (this will vary between door sizes, whether they are imperial or metric sized. So beware of the term 'Standard Opening Size' – *ALWAYS CHECK BEFORE ORDERING*).
(b) The width of the architrave (generally available from 50 to 75 mm EX) – see Fig. 8.5.
(c) The condition of the architrave. This is very important – use only joinery quality. Reject any lengths with severe defects, such as large 'dead knots', or those which are 'sprung' – see Volume 1, section 1.7.8 (Table 1.7).

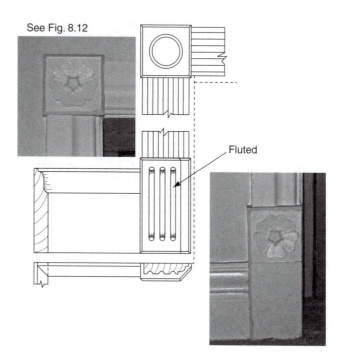

See Fig. 8.12

Fluted

Fig. 8.13 *Incorporating decorative features within corner and plinth blocks*

Timber merchants can supply architrave as 'sets' (2 legs, 1 head) with a small provision of waste – again you must state the door's size.

For example, an imperial door 6' 6'' × 2' 6'' (1981 mm × 762 mm) using 50 mm wide architrave would require:

2 legs @ 2.1 m
1 head @ 0.9 m

and would require a standard 5.1 m length.

Beware – wider architraves would require a longer head length and possibly a longer leg length.

8.2.2 Fixing architraves (three-sided openings)

Figure 8.14 shows two methods and sequences of fixing architrave around a rectangular door opening with mitred head and dropped legs.

METHOD 'A'
STAGE 1 – First leg

(a) Mark the margin distance – usually about 4–6 mm at intervals around the edge of the door lining. The thickness of one leg of a 4-fold joiner's rule (Volume 1, section 5.1.2) or combination square (Volume 1, Fig. 5.9b) is usually adequate (narrow margins with non-tapering architraves can restrict the amount the door is allowed to swing open without catching it. Narrow margins can also restrict the knuckle of the hinge housing and the lead-in of a lock striking-plate (see Fig. 4.82). Large margins on the other hand reduce the width required for fixing, particularly with rebated casings.

Where several doors are involved it may be worth considering using a 'margin template' like the one shown in Fig. 8.15. It can be made from either solid wood (preferably dense hardwood), plywood, or MDF with a rebate formed along two of its edges to the required depth to suit the margin width and shoulder depth to allow it to sit up against the door lining.

(b) Position the leg of the architrave against the wall as shown and mark the point where the internal margin for the leg meets the head margin.

(c) This is the 'heel' of the mitre, which will be cut at 45° from the 'toe'.

(d) With the aid of a 'mitre box' (Volume 1, Fig. 5.96k) and a fine-toothed saw cut the mitre – avoid using a 'mitre block' (Volume 1, Fig. 5.96j) as this can be less accurate. Alternatively, use a proprietary hand mitre saw (Volume 1, Fig. 5.97), or electric 'crosscut mitre saw' (Volume 1, Fig. 6.29).

(e) Using 38–45 mm oval nails fix (leaving nail heads to draw) to the door lining.

Note: *If there is any doubt as to the squareness of the opening, use a short end of architrave as a gauge to pencil in cross lines as shown at each corner. This will allow you to mark a true bisected mitre for an accurate heel and toe angle. Any minor adjustments to the mitre joint can be made by taking fine shavings off with a block plane (see Volume 1, section 5.4.2).*

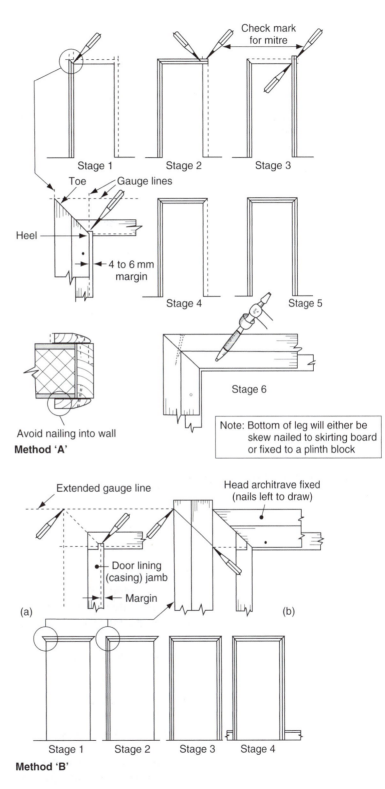

Fig. 8.14 *Two methods of fixing architrave around a door opening (see also Fig. 4.67)*

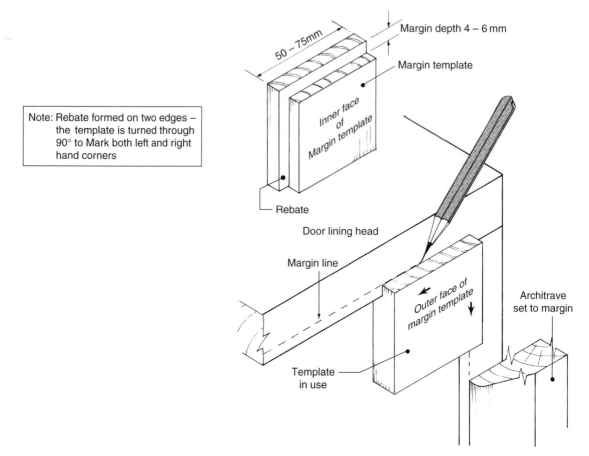

Note: Rebate formed on two edges –
the template is turned through
90° to Mark both left and right
hand corners

Fig. 8.15 *Architrave margin template*

STAGE 2 – Head architrave

(a) Mitre the adjoining end of the head architrave – adjust by planing if required.
(b) Once fitted mark the position of the heel to toe at the other end.
(c) Cut the mitre *but do not fix in place.*

STAGE 3 – Second leg

(a) Position and mark the top mitre as before.
(b) Cut the mitre.

STAGE 4 – Head architrave

(a) To the margin lines fix to the door lining using 38–45 mm oval
nails (leaving nail heads to draw).

STAGE 5 – Second leg

(a) Reposition architrave leg – adjust by planing if necessary.
(b) To the margin lines fix to the door lining using 38–45 mm oval nails (leaving nail heads to draw).

STAGE 6 – Final stage

(a) Complete the nailing process at about 300 mm centres.
(b) Ensuring that the outer edges of the mitre are in line, from the top of the mitre drive a nail at an angle across the mitre as shown (Fig. 8.14).
(c) Punch all the nails just below the surface.
(d) The bottom outer edge will eventually be secured by skew-nailing to the skirting board.

METHOD 'B'
STAGE 1 – Head architrave

(a) Mark the margin distance – as in method 'A'.
(b) As shown at 'a' (Fig. 8.14) – using a short end of architrave mark cross-over gauge lines at each corner.
(c) Cut a mitre at one end – as in method 'A'.
(d) Whilst holding the head architrave in position against the gauge lines on the lining and plasterwork check the heel and toe of the mitre for fit – adjust as necessary.
(e) Whilst holding the head architrave in position mark the heel and toe of the opposite end – check for 45° and cut as before.
(f) Fix (leaving nail heads to draw) to the door lining – as method 'A'.

STAGE 2 – First architrave leg

(a) Hold plumb against the head architrave – mark the toe and heel from the extended head gauge lines as shown at 'a' (Fig. 8.14).
(b) Check the bisection line with your mitre or combination try square – if satisfactory cut as before using a mitre box or 'mitre-saw' (see Volume 1, section 6.11).
(c) If it is found to be off 45°, cut by hand and if required fit to head mitre using a block plane.
(d) Fix (leaving nail heads to draw) to the door lining – as method 'A'.

STAGE 3 – Second leg
Repeat as Stage 2.

STAGE 4 – Final stage – as method 'A'.

8.2.3 Scribing architraves

Where an architrave has to abut an adjacent wall, and it spans a gap narrower than its width and its abutting surface is either uneven or out of plumb, it will require shaping to ensure a close fit against its abutment. Architrave around a rectangular door opening with mitred head and dropped legs (Fig. 8.16).

Figure 8.16a shows a situation which will require both one leg and part of its head scribing. Figure 8.16b shows the finished arrangement; collectively Fig. 8.16a–d show methods of achieving this, for example:

(a)   Cut the left architrave to full height 'h' (Fig. 8.16a).
(b)   Determine the scribing width 'x' (Fig. 8.16c) by measuring that portion which overhangs the lining when held plumb against the wall, plus the margin distance.
(c)   Either cut a scribing block (or use dividers or a compass) to width 'x' (Fig. 8.16c).
(d)   Temporarily tack the architrave plumb to a distance as shown in Fig. 8.16c then mark the scribe line down the full length of the architrave.
(e)   Detach the scribed leg.
(f)   Remove the waste wood either by planing or sawing (Fig. 8.16d) – slightly undercut as shown in Fig. 8.16d to ensure a good fit.
(g)   Reposition, check the fit, then mark and cut the mitre.
(h)   Cut the head architrave to length 'w' plus overhang (Fig. 8.16a).
(i)   Position the head architrave at a distance as shown in Fig. 8.16c – temporarily tack in place.
(j)   Scribe to the bulkhead ceiling line as previously described above.

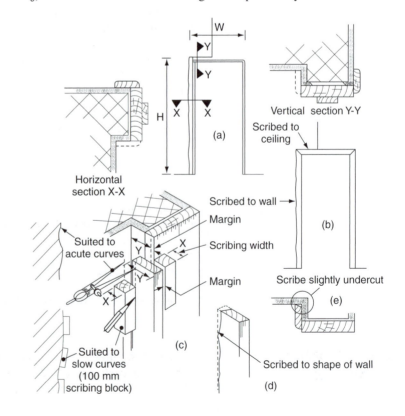

Fig. 8.16 *Scribing architrave to an uneven surface of a wall or ceiling*

(k) Remove the waste wood either by planing or sawing – slightly undercut as shown in Fig. 8.16d to ensure a good fit.

(l) Reposition, check the fit, then mark and cut the mitre.

(m) Repeat the procedures as described in section 8.2.2.

8.2.4 Fixing plinth blocks (also known as 'Architrave block', see also Fig. 8.13)

As shown in Fig. 8.17a, where the architrave is thicker than the skirting board there is, unless the lower corner of the architrave needs protection, no need for a plinth block. If, on the other hand, the skirting board is thicker than the architrave, and/or protection is required (Fig. 8.17b), a plinth block is usually the answer (Fig. 8.17c).

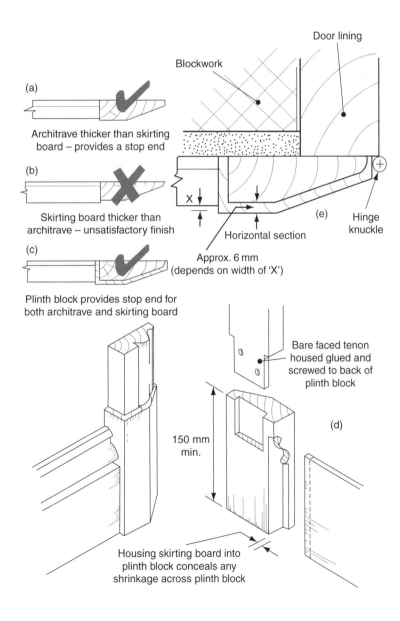

Fig. 8.17 *Using a plinth block*

As shown in Fig. 8.17d, plinth blocks are traditionally first fixed to the architrave via a bare faced tenon glue and screwed to the back of the block – the legs are fixed in the normal way. The block will then form an end stop for the skirting board, the profile of which is usually housed into the block. In this way any shrinkage gaps that may occur are shrouded.

Alternatively, both the architrave and the skirting board can be butted against the block. In this way the block is fixed first, followed by the architrave, then the skirting board.

The relationship between the architrave and the plinth block can vary. An example is shown in the horizontal section in Fig. 8.17e.

8.2.5 Fixing architraves (four-sided openings)

Figure 8.18 shows a method and sequence of fixing architraves around a wall serving hatch or ceiling access trap.

STAGE 1 – Marking out

(a) Mark a margin distance of about 3–5 mm around the opening;
(b) If there is any doubt as to the squareness of the opening (pay particular attention to ceiling openings), use a short end of architrave to gauge the width at each corner to enable the true mitre angle to be formed.

> Note: *This procedure can be used to find the mitre angle of any angle (both acute and obtuse)*

STAGE 2 – Left-hand architrave (a).

(a) Cut the bottom mitre;
(b) Hold firmly into position whilst marking the heal point of the top mitre;
(c) Nail into position – leave nail heads to draw.

STAGE 3 – Head architrave (b)

(a) Cut and fit the left-hand mitre;
(b) Hold firmly into position whilst marking the heal point of the RH mitre – cut the mitre and put to one side.

STAGE 4 – Bottom architrave (c)

(a) Cut and fit left-hand mitre;
(b) Hold firmly into position whilst marking the heal point of the right-hand mitre – cut the mitre.
(c) Nail into position – leave nail heads to draw.

STAGE 5 – Right-hand architrave (d)

(a) Cut and fit bottom mitre;
(b) Hold firmly into position whilst marking the heal point of the top mitre – cut the mitre.

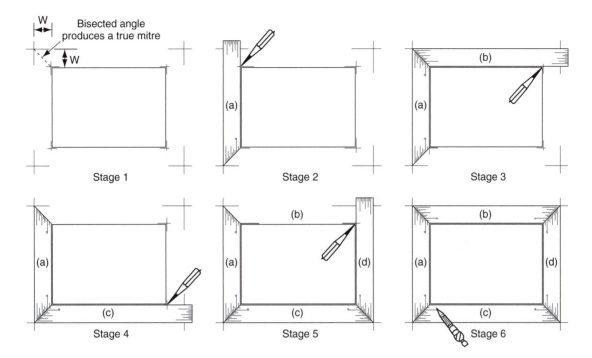

Fig. 8.18 *Sequence of fixing architraves around wall and ceiling openings*

Note: *Figure 8.14 shows architrave around a rectangular door opening with mitred head and dropped legs.*

STAGE 6 – Final assembly

(a) Hold (b) and (d) in position – check mitres (a–b, b–d, c–d) for fit – adjust if required;
(b) Nail 'd' into position, then 'b';
(c) Ensuring that all the mitres are fair-faced, carefully nail across each mitre;
(d) Complete the nailing pattern then punch all nail heads just below the surface in preparation to receive any filler;
(e) Using a block plane remove a small arris (sharp corner) all round the framework.

8.3 Skirting boards

We have already seen how skirting boards form a decorative finish between the wall and floor, and how they offer protection to the wall against everyday activities like vacuuming the carpet or sweeping the floor. The skirting board size (in particularly its depth) and moulding details usually reflect the period in which the property was built.

A selection of commonly available moulded sections is shown in Fig. 8.19 (also see Fig. 8.5). If wood is to be used, available lengths are as for processed timber, for example, lengths will start at 1.8 m then upwards in increments (multiples) of 300 mm up to a maximum of 6.3 m.

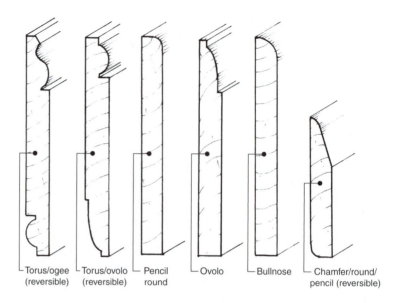

Fig. 8.19 *Commonly available moulded sections used as skirting board*

Torus/ogee (reversible)   Torus/ovolo (reversible)   Pencil round   Ovolo   Bullnose   Chamfer/round/pencil (reversible)

8.3.1 Corner joints

External and internal corners are dealt with differently. External corners are mitred (Fig. 8.20), internal corners are scribed (Fig. 8.21). The main reason for using a scribed joint is to minimize the shrinkage gap across the width of the skirting board; remember that the greatest amount of shrinkage across a section of timber takes place tangentially, followed by radially and then finally the least is longitudinally (see Volume 1, section 1.7.2–3). Figure 8.22 shows how this can affect an internal mitre. The amount of shrinkage will depend on the moisture content of the wood on installation and the position of the growth rings, for example are they predominantly across the width as with quarter sawn timber, or down its depth as with tangentially sawn timber (see Volume 1, section 1.7.3).

8.3.1.1 The mitre

Very few corners are perfectly square, but just like those angles which are acute (less than 90 degrees) or obtuse (greater than 90 degrees) the mitres angle, as shown in Fig. 8.23 can be easily found, by bisection of the angle. This is best done by using a mitre bevel board, which needs to be between 75mm to 100mm wide and about 900mm in length.

The following method of bisecting any angle can be used to great effect:

1. Position the mitre board flat on the floor and up against one wall, overlapping the corner by as much as its own width – mark off the floor at its outer edges.
2. Reposition the board to touch the other wall and mark floor as before.
3. By marking a line from the corner point of the wall to where the outer lines intersect, the bisecting line will produce a true mitre line.
4. The mitre line can now be marked onto a bevel board ready to be transferred to a sliding bevel.

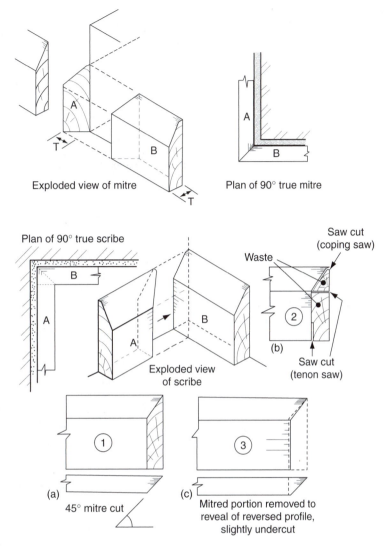

Fig. 8.20 *Cutting a true mitre to a 90° external corner*

Fig. 8.21 *Forming a true scribe (reverse profile) to a 90° internal corner*

5. This angle can be transferred direct to the skirting board, or used as the angle for marking onto a mitre box.

8.3.1.2 The scribe

As we have seen, internal corners are scribed to reduce the effect of shrink as can be seen from Fig. 8.21. Cutting a scribed joint is not as straightforward as using a mitre. It involves several processes, for example:

STAGE 1
Using a mitre box cut a true mitre as if to fit an internal corner. This in effect will produce a front view of the skirting boards profile as shown in Fig. 8.21a.

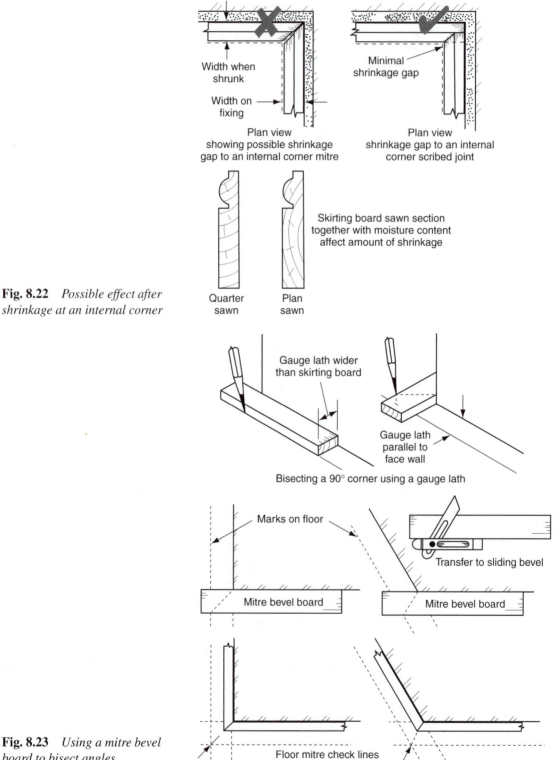

Fig. 8.22 *Possible effect after shrinkage at an internal corner*

Fig. 8.23 *Using a mitre bevel board to bisect angles*

STAGE 2
Remove the waste portion (Fig. 8.21b and 8.21c). With the exception of the top cut, cuts may be slightly undercut – this will ensure a close-fitting joint.

STAGE 3
Before fixing offer the scribe up to the face of a short end of skirting board to check to check the fit.

Note: *This method of cutting a reverse profile can be used on any moulded profile; curved sections are generally cut away with the use of a coping saw.*

8.3.2 Stop ends

Whenever a moulding comes to a stop without an abutment its mould is either returned to the wall it is fixed to or the floor it sits on.

8.3.2.1 *Returned to wall*

Fig. 8.24 shows two methods of returning the moulded profile. In Fig. 8.24a the profile is cut across the end grain – this method is known as a Mason's mitre. The second method shown in Fig. 8.24b is used with complex moulds and quality work: a true mitre is formed with a short end of skirting board, which is glued and pinned together. When set, the waste end is carefully cut back leaving a true return without any end-grain exposure.

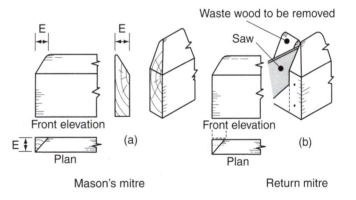

Fig. 8.24 *Returning the end profile of a skirting board towards a wall*

8.3.2.2 *Returned to floor*

Fig. 8.25 shows two methods of returning a moulded profile back towards the floor. The first example (Fig. 8.25a) is only suitable for a skirting board with a chamfered face, whereas Fig. 8.25b can be used with any shape of mould – even the most complex.

8.3.3 Lengthening skirting board (heading joints)

Heading joints are only used where the run is too great for a standard length of skirting board – with good planning heading joints can often the avoided.

If such a joint is required, as shown in Fig. 8.26, it is formed by making two accurate opposing 45° cuts using a mitre box. Great care is

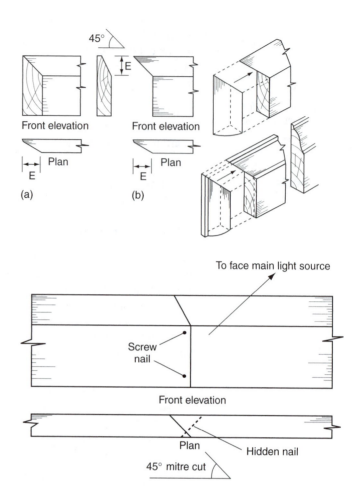

Fig. 8.25 *Returning the end profile of a skirting board back towards a floor*

Fig. 8.26 *A lengthening or heading joint*

required for this joint as any inaccuracy will be difficult to conceal. The overlapping portion should face the main natural light source (window) to reduce the effect of any shadow line. The joint is secured by skew nailing.

8.3.4 Fixing skirting boards

Before any fixing takes place it is important to plan the sequence you intend to follow – not only for economic reasons but to avoid cutting a mitre and/or scribe at both ends. Two fixing sequences are shown in Fig. 8.27.

In Fig. 8.27a the sequence follows a clockwise direction, whereas Fig. 8.27b follows an anti-clockwise direction – only lengths numbered three, seven, and nine have either a mitre or scribe at both ends.

Skirting boards are generally fixed to walls by nailing through the face of a board. As shown in Table 8.2 the methods used can vary according

Note: *Always ensure that any concealed services (electric cables, etc.) are detected (see section 8.9) before any fixings are made*

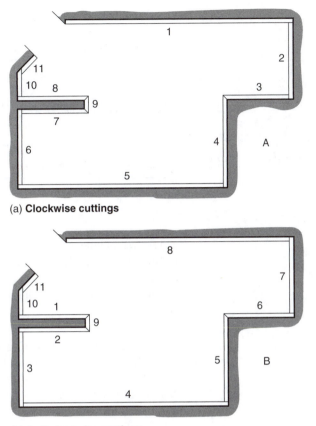

Fig. 8.27 *Possible sequences for arranging and fixing skirting board*

(a) **Clockwise cuttings**

(b) **Anti-clockwise cuttings**

Table 8.2 Wall trims and finishes (mouldings) in relation to their use

Wall base material	Fixing medium	Fixing	Comments
Brickwork	Plastic plugs	Screws	Various Proprietary makes
Brickwork	Direct	† Masonry nails	Goggles must be worn
Breeze (clinker) blocks	Direct	† Masonry nails	Goggles must be worn
Breeze (clinker) blocks	Plastic plugs	Screws	Various Proprietary makes
Aerated concrete blocks	Direct	Cut clasp nails	Not readily available
Aerated concrete blocks	Special Plastic plugs	Screws	Various Proprietary makes
Most base materials	Special Wall Adhesive	Adhesion	Various Proprietary makes

Note: As a general rule the length of the fixing is usually 2½ × the thickness of the material to be fixed.
† Masonry nails should penetrate the base material by not less than 25 mm.

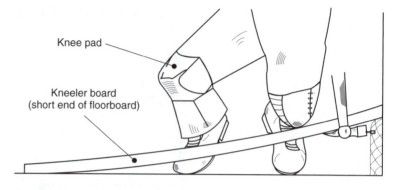

Knee pad

Kneeler board
(short end of floorboard)

Fig. 8.28   *Adding pressure to a 'sprung' length of skirting board prior to fixing*

to the different types of base material used for walls (see also Volume 1, section 12, Fixings devices).

Fixing centres can vary – with long lengths 600 mm centres are usually adequate.

More often than not the skirting board will be slightly sprung in its length. It will therefore need to be pushed to the floor before being fixed. As shown in Fig. 8.28 a short end of board (kneeler board) can be used for this purpose.

8.3.5 Scribing skirting boards

If, as shown in Fig. 8.29a, the floor is uneven and a close fit is required then it can be scribed similar to architrave. However, a word of warning – you should always bear in mind that if you reduce the depth of the skirting board at the ends (Fig. 8.29b and 8.29c), the lengths of the abutting skirting boards should be reduced to bring them in line. The alternative is to increase the depth of the skirting board to be scribed by gluing a strip of timber to the under side (Fig 8.29c). Then, by using an offcut of the strip as a scribing block, scribe the skirting board to the floor.

8.4 Rails around a room

As shown in Table 8.1 the two rails commonly fixed around a room are the 'dado' and 'picture rail'. The dado rail can have two main functions: firstly to divide the height of the room into two horizontal areas, and secondly as a means of protecting the wall from being damaged from chair backs and wall furniture, hence the reason for calling it a '*chair wall rail*', or possibly a '*waist rail*' because it is usually situated at about waist height. Figure 8.30 shows four typical dado sections.

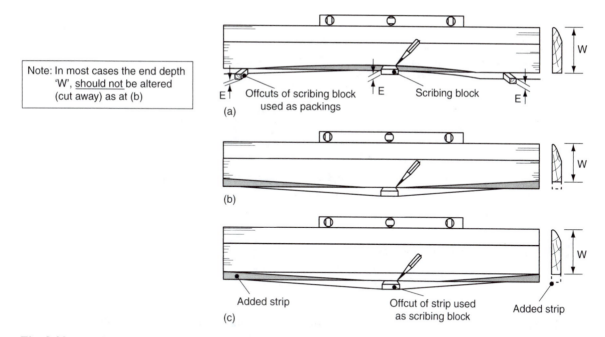

Note: In most cases the end depth 'W', <u>should not</u> be altered (cut away) as at (b)

Fig. 8.29 *Scribing skirting board to an uneven floor*

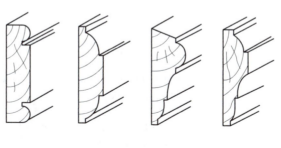

Fig. 8.30 *Examples of shaped sections used as dado rails*

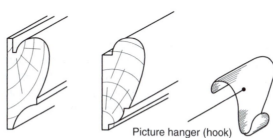

Fig. 8.31 *Examples of shaped sections used as picture rails*

Picture hanger (hook)

Originally, the primary function of the picture rail (Fig. 8.31) was to act as a bracket for a picture hook from which a picture was hung – these could be very large and heavy so their fixings were very important. Today, picture rails, just like dado rails, are often provided more as a decorative feature.

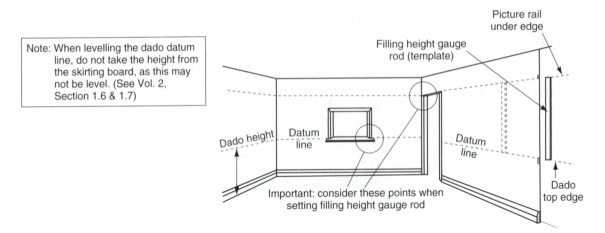

Fig. 8.32 *Setting out walls of a room for fixing dado and picture rails*

8.4.1 Setting out

Figure 8.32 illustrates the importance of taking a datum line around the room. There are several methods of striking this line (see Volume 2, section 1.7) including:

(a) Traditional spirit levels;
(b) Digital spirit levels;
(c) Water levels;
(d) 'Cowley' automatic level;
(e) Laser levels.

Once this level line is struck – ideally as the top line of the dado rail – this can be repeated at the picture rail height. It can be easier to use a 'gauge rod' cut to the filling height – this should be kept and marked 'template' as it will be useful at a later stage when, assuming that a dado rail is to be fixed first, the picture rail is fixed to the wall.

8.4.2 Treatment at corners (see also section 8.8, Raking moulds)

The same principles as those used for the skirting board will apply. When corners have to be bisected, the methods shown in Fig. 8.23 can still be used, but if there are wall units covering the floor at this point use two short ends of board hand held across the corner instead of laying them on the floor.

8.4.3 Stop ends

The same methods as those shown in Fig. 8.24 can be used. If, however, a complicated shape is to be returned in the solid then, as shown in Fig. 8.33, the profile can be marked from the back. Then very carefully cut away the waste wood with a coping saw; with its blade set to cut on the pull stroke, ragged edges on the face mould can be avoided. The shape can then be finished with a sharp chisel and abrasive paper held round the appropriate former.

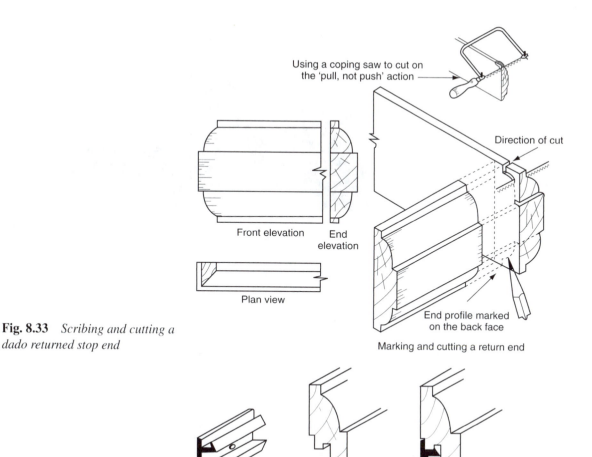

Using a coping saw to cut on the 'pull, not push' action

Direction of cut

Front elevation

End elevation

Plan view

End profile marked on the back face

Marking and cutting a return end

Fig. 8.33 *Scribing and cutting a dado returned stop end*

Fig. 8.34 *Proprietary concealed fixing device for fixing dado rails*

8.4.4 Lengthening dado and picture rails (heading joints) (see also section 8.3.3)

Lengthening joints should always be avoided; the picture rail should never be joined in its length. Dado rails can be joined by using adjacent 45° cuts as shown in Fig. 8.26.

8.4.5 Fixing rails

Note: *Special regard should be given to fixing picture rails as the occupants could decide to use them as they were originally intended, i.e. for hanging pictures and mirrors from. These could be extremely heavy so it is important to take this into account at the fixing stage.*

The sequence described in section 8.3.4 for fixing skirting boards will still apply, as well as the types of fixing used as shown in Table 8.2. Concealed proprietary snap fixing suitable for dado rails is available – a typical example is shown in Fig. 8.34 (they should be fixed according to the manufacturer's instructions).

Fig. 8.35   *Example of the use of a Delft shelf*

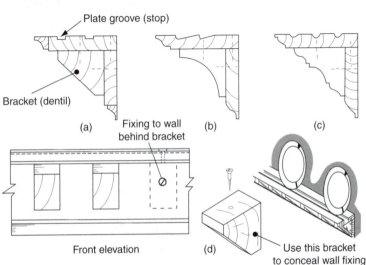

Fig. 8.36   *Different Delft shelf designs*

8.5 Delft (plate) shelf (Fig. 8.35)

These shelves were originally designed to display 'Delftware' plates, which typically have a blue decoration on a white background. They are usually associated with hallways or dining rooms where they support ornamental plates or ornaments to good effect, out of harm's way. They can very often take the place of a picture rail.

Figure 8.36a shows a modern design, whereas Fig. 8.36b and 8.36c illustrate two classical designs – all have dentil bracket supports. Sizes range from 100 mm by 100 mm to 150 mm by 150 mm; they are usually fixed to the wall by using screw and plugs. Figure 8.36d shows how the fixings can be concealed behind a bracket.

8.6 Cornice (drop or false cornice)

Cornices are more usually associated with plasterwork where wall meets ceiling, or the finish around the top of a wall or cupboard unit. There are, however, occasions were we can use a cornice to head wall panelling or independently as a feature, and as a means of housing concealed lighting to illuminate the freeze area (Fig. 8.1) of a room.

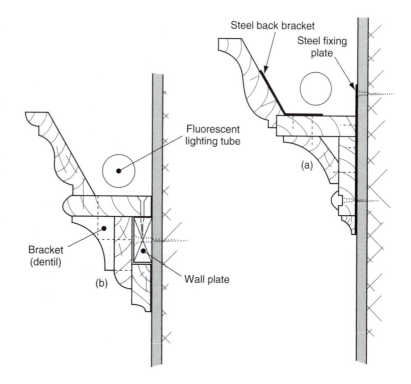

Steel back bracket

Steel fixing plate

Fluorescent lighting tube

(a)

Bracket (dentil)

Wall plate

(b)

Fig. 8.37 *Wood cornice with concealed lighting*

Figure 8.37 shows examples of cornice arrangements capable of concealing fluorescent tubes. The illustration in Fig. 8.37a is made up of two runs of moulding and a shelf, whereas Fig. 8.37b is built up of a wall plate with a horizontal board, dentil brackets and moulded capping boards.

8.7 Infill panel frame

This type of framework is purely decorative. Traditionally this work was undertaken by the plasterer using a fibrous plasterwork technique (a skill still used today). However, a less expensive alternative is to use a wood moulding. Figure 8.38 shows examples of the type of profile that could be used for this purpose. There is a great variety in proprietary decorative mouldings on the market, which also include those which have been embossed or even carved.

The infill panel arrangements can vary greatly in their design, usually to suit the decor of the room. Figure 8.39 shows how corner details may vary. In Fig. 8.39a a basic 45° mitre cut is made, whereas in Fig. 8.39b the mitre is formed by bisecting the angle as shown in Fig. 8.21. The mitre for the quadrant at Fig. 8.39c is more complex. In this case, as shown in Fig. 8.39d, the mitre line is developed using geometry to produce a curved mitre line. Had a straight cut been made, the abutting moulding detailing would not have met up as shown in Fig. 8.39e. Alternatively, as shown in Fig. 8.39f, the curved moulding is widened to meet the requirements at 'X'.

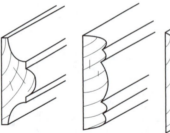

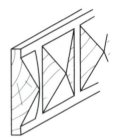

Machine-cut moulds   Embossed   Machine carved
                            (pressed) mould

Fig. 8.38 *Available mouldings for constructing infill panels*

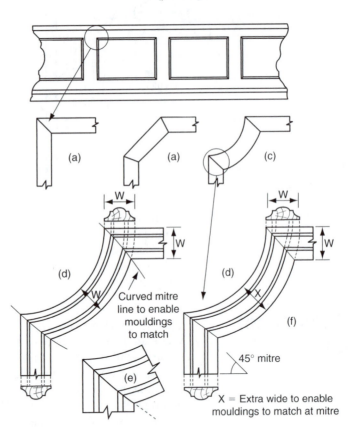

Fig. 8.39 *Corner details for an infill panel*

8.7.1 Fixing panel mouldings

Methods of fixing the moulding to the wall really depend on their section size, length and condition (i.e. if they have any slight distortions). For example, the light sections of moulding may be stuck to the wall using a suitable panel adhesive. Heavier or longer sections will require mechanical fixings (Table 8.2).

8.8 Raking moulds (Fig. 8.40)

Moulding trims inclined at an angle (raking moulds) are used to eventually join up to one or more different levels to provide continuity of the moulding. In this case we are concerned with raking moulds which continue around a corner. Three examples are given below and shown in Fig. 8.40.

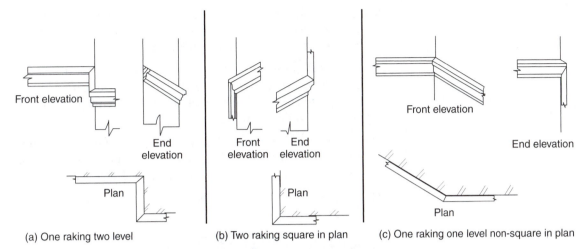

(a) One raking two level

(b) Two raking square in plan

(c) One raking one level non-square in plan

Fig. 8.40 *Possible arrangements for raking moulding*

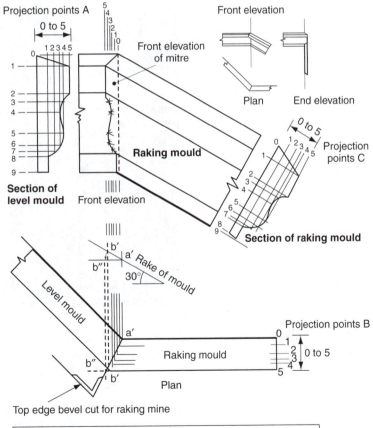

Fig. 8.41 *Raking moulding – bevel and section details*

Note: Projection points of A, B and C are the same distance apart

(a) One raking and two level – square in plan (Fig. 8.40a);
(b) Two raking – square in plan (Fig. 8.40b);
(c) One raking one level – non square in plan (Fig. 8.40c).

In these situations it would not be advisable, or indeed practicable, to use the same section timber for both the level and inclined mould, if the moulds are to correctly meet at the corner mitres.

Figure 8.41 illustrates a method of finding the bevel for the back of the moulding, the top edge bevel for the inclined mitre, and the required profile of the inclined moulding if the overall profile detailing is to meet at the mitre joint.

Figure 8.42 shows a method of using the same section mould, either level or inclined throughout, by inserting a small level mould before and/or after each corner mitre – internal scribed, external mitred.

The face mitres for the level and inclined moulds are formed by bisecting the angles (see Fig. 8.23).

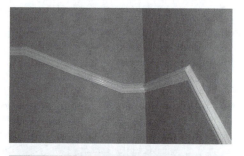

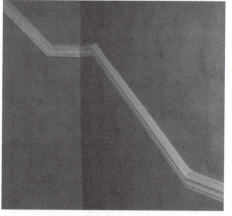

Fig. 8.42 *Raking moulding* in situ

8.9 Detection of services

Before any refurbishment or new work is carried out, carpenters and joiners must always be aware of the possibility of any hidden services, particularly electrical ones which may be hidden behind existing partitions or solid walls – coming into contact with live cables while fixing could be *fatal*.

However, there are various battery operated devices on the market that are capable or detecting electrical cables, pipework and timber studs. One example is shown in Fig. 8.43.

Using the various modes settings, it is capable of detecting and locating any brickwork, concrete and wooden structures before drilling work commences.

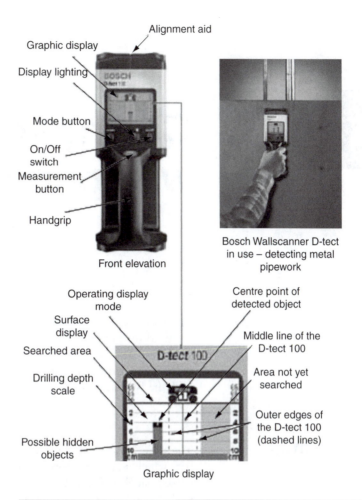

Fig. 8.43 *Bosch Wallscanner D-Tect 100*

References

Health and Safety at Work Act (HSAWA) 1974

The Construction (Health, Safety and Welfare) Regulations 1996

BS EN 923: Adhesives. Terms and definitions.

BS EN 1186–2: 1988, Timber for, and workmanship in joinery. Specification for workmanship.

BS 8417: 2003, Preservation of timber – recommendations.

BS 5973: 1993, Code of practice for access and working scaffolds and special scaffold structures in steel.

BS 1139–3: 1994, Specification for pre-fabricated access and working towers.

BS 1139–4: 1982, Specification for pre-fabricated steel splitheads and trestles.

BS 1129: 1990, Timber ladders, steps, trestles, and lightweight stagings.

BS 2037: 1994, Aluminium ladders, steps, and trestles for the building and civil engineering industries.

BS 2482: 1981, Specification for timber scaffold boards.

BS 5973: 1993, Section 9 – tying to buildings.

9

Casing-in and wall panelling

In relation to the internal finishes of a building, these areas of work are used for practical and economical reasons to cover or resurface structural and non-structural items, which either are not in keeping with their surroundings and/or provide fire protection in accordance with the Building Regulations.

9.1 Casings

The term 'casings' usually refers to the framework, facework, and trim which go towards forming an enclosure into which service pipes, cables, steel and concrete beams and columns, etc., are or could be housed. Figure 9.1 shows examples of where casings are often found.

9.1.1 Steel beams (binders)

These steel beams – often referred to as 'binders' – are used within intermediate framed floors to reduce the effective span of bridging joists, just as purlins are used within the roof structure to reduce the span of roof rafters. A floor using one of these beams would be known as a 'double floor', one using two beams as a 'triple floor', and so on.

Structural steelwork will have to be clad to protect it from the effects of fire – the amount of fire resistance will depend on the location and function.

Suitable materials which contribute to fire resistance include concrete, plaster (coated plasterboard), and non-combustible boards. Board materials will require some form of groundwork onto which they can be fastened.

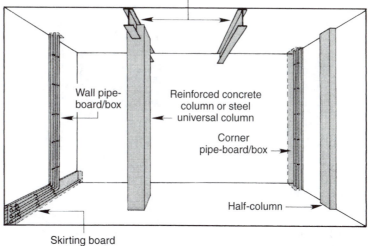

Fig. 9.1 *Examples of situations requiring casing (casings)*

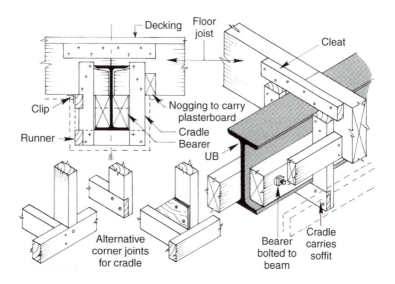

Fig. 9.2 *Timber cradles and grounds for casing a partially exposed steel beam*

Figures 9.2 and 9.3 show how timber groundwork can be formed and fixed around a steel universal beam (UB) to receive a fire-protective boarding.

Encasement support (Fig. 9.2) will require either:

(a) A series of cradles fixed at centres of 450 to 600 mm, depending on the sectional size of any counter battens or panel thickness;
(b) Framed grounds to both sides of the beam and its underside.

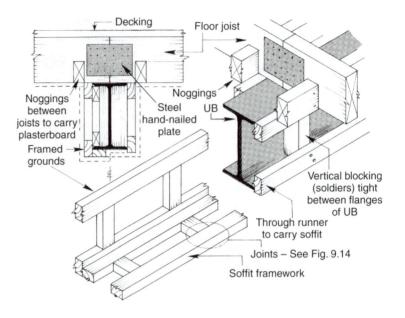

Fig. 9.3 *Timber cradles and grounds for casing a fully exposed steel beam*

Both these means of support will require fixing points. These may be provided by:

(a) Joist bearers which would have been installed to support the lower edge of the floor joists – these may be connected to the beams via a bolt set through the web of the beam;

(b) Vertical blocking (soldiers) – these should fit tight between the flanges of the beam.

9.1.1.1 Cradles

Consist of a series of 'U' shaped prefabricated frames constructed by using any suitable framing joint (Fig. 9.14). The top of each of the cradle legs is fixed to the sides of every joist, and the lower portion fixed to bearers.

9.1.1.2 Framed grounds

Consist of a light prefabricated timber framework using a choice of framing joints (Fig. 9.14). Side grounds will be attached to the vertical blocking; soffit grounds will be fixed to the side grounds.

9.1.1.3 Unframed grounds

Consist of an assembly of timber runners attached to vertical blocking.

9.1.1.4 Facings

Finally the grounds will be clad with a fire-resistant material that must be within the accepted limits of the '*spread of flame*' class to comply with current Building Regulations.

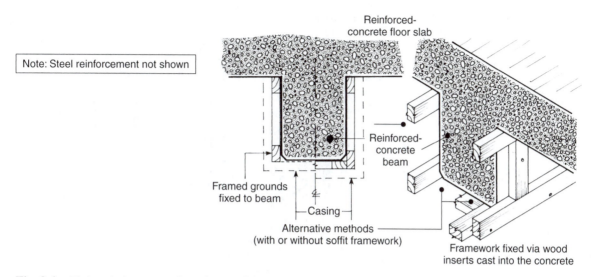

Note: Steel reinforcement not shown

Reinforced-concrete floor slab

Reinforced-concrete beam

Framed grounds fixed to beam

Casing

Alternative methods (with or without soffit framework)

Framework fixed via wood inserts cast into the concrete

Fig. 9.4   *Fixing timber grounds with or without soffit to a concrete beam*

9.1.2 Concrete beams

In this case, before any grounds are assembled it is good to check for any discrepancies in beam size. Timber grounds will be attached to the beam using a method approved by the building designer – this may involve using a cartridge operated fixing tool (Volume 1, section 8) to drive the hardened steel pins into the concrete.

Figure 9.4 shows an optional arrangement for fixing the grounds. Either:

(a)   Two side frames and an under beam frame would be used; or
(b)   The under beam framework has been dispensed with to allow the facing soffit material to bridge between them. In this case extra care must be taken to ensure that the facing materials are in line and the soffit width is uniform along its length.

Groundwork can then be clad to enhance the beam's appearance or to match with a ceiling, providing it meets and complies with current Building Regulations in relation to surface spread of flame, etc.

9.1.3 Steel columns

These can be dealt with in much the same way. With the exception of three-sided half-columns (divided by a wall), timber groundwork can be self-restraining. Figure 9.5 shows two alternative ways of fixing timber grounds to a steel column. If provision has been made for bolting timber blocks between the flanges, groundwork may be necessary on only two sides (Fig. 9.5a), otherwise blocks can be trapped between the flanges (Fig. 9.5b) to act as fixing and stabilizing points for the groundwork, which envelopes the whole column as shown.

9.1.4 Concrete columns

Casing a concrete column is shown in Fig. 9.6. A set of four framed grounds is made-up so that, when they are together as shown, their

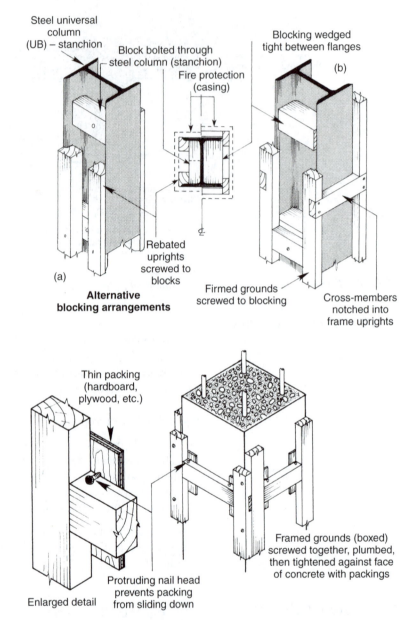

Steel universal column (UB) – stanchion

Block bolted through steel column (stanchion)

Fire protection (casing)

Blocking wedged tight between flanges

(b)

Rebated uprights screwed to blocks

(a)

Alternative blocking arrangements

Firmed grounds screwed to blocking

Cross-members notched into frame uprights

Fig. 9.5   *Fixing timber ground around steel columns*

Thin packing (hardboard, plywood, etc.)

Framed grounds (boxed) screwed together, plumbed, then tightened against face of concrete with packings

Protruding nail head prevents packing from sliding down

Enlarged detail

Fig. 9.6   *Fixing timber grounds around a concrete column*

interconnecting internal dimensions are slightly larger than the column. Before proceeding, however, checks must be made to ascertain the largest cross-section of the column to ensure that the framework will be slightly oversize at all points. Slight adjustments can then be made by using stopped packings between the face of the column and the backs of the grounds. These packings must be prevented from falling down behind the grounds, if, or when more likely, shrinkage takes place due to moisture movement. Therefore, as a precaution, a nail can be driven into the packing – just above the cross member as shown in Fig. 9.6.

9.1.5 Fire resistance

Steelwork, although non-combustible will lose structural strength and distort in temperatures at about 800°C. Steelwork must therefore be protected from heat generated by fire for a specific period of time to enable a compartment or building as a whole to be safely evacuated.

To achieve this integrity structural steelwork must be able to attain a prescribed period of fire resistance. This may be obtained either by spraying with a fire resistant coating and/or being encased with a non-combustible material. Materials such as gypsum-based plaster-board, and many other proprietary products manufactured specifically for this purpose, could be used.

9.1.6 Casings for service pipes and cables

In the main, pipes and cables are hidden from view – provision for them having been made within a hollow floor or in purpose-made ducts in both the floor and the walls. There are, however, situations where such services have to be brought out on to the surface, and it is in these situations that provisions such as those shown in Figs 9.7 and 9.8 can be applied.

Figure 9.7a shows how a pipe-box can be formed behind a skirting board. It is always an advantage if the front is detachable, either fully or over the greater part of its length – access is then possible without undue damage to decorations if maintenance or 'add-ons' are required. Where edge joints occur 'veeing' not only makes them less obvious but also aids dismantling – a trimming-knife (with the blade retracted to its minimum) can simply be run along the joint to cut the film of paint or varnish. By under bevelling the edge or edges of the pipe-board, the joint between the wood and the plaster is keyed and is less apparent.

Figure 9.7b and 9.7c deal with situations where pipes or cables are sited vertically. Provided the pipe-boards have been fixed to the walls at the first fixing stage of construction, forming a detachable casing as shown is relatively easy. However, before front covers (casings) are fixed, it is important to ensure that where the pipes, etc. pass through the ceiling and/or floor, this area is suitably sealed to form a cavity barrier (fire stopped), thereby preventing the passage of flame. Otherwise, in the event of a fire, gaps could let the ducting act as a flue, allowing fire to spread rapidly between floors.

Where a backboard (pipe-board) is not provided at the first fix stage, the arrangements shown can be modified by fixing timber grounds directly to the walls as shown in Fig. 9.8.

Front boards (inspection covers) should be screwed into position – the use of screw cups and/or domes (see Volume 1, Table 12.6) is recommended.

Important Note: *Where services pass through a ceiling or wall of a fire-resistant construction any gaps or openings around these pipes or cables, etc. must be properly fire stopped and sealed before they are encased or hidden*

Note: Pipe-board fixed to wall at first-fixing stage of building construction. All ducts must be sealed where they meet walls, floors and ceiling

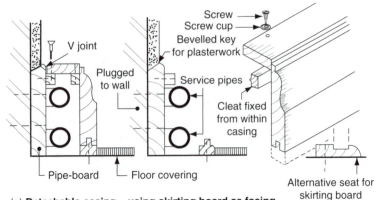

(a) **Detachable casing – using skirting board as facing**

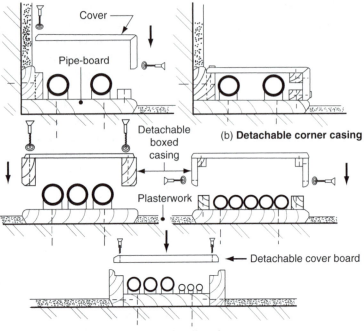

(b) **Detachable corner casing**

(c) **Detachable face of wall casings**

Fig. 9.7 *Casing around service pipes and cables*

In situations where a detachable facing is not practicable for fear of disturbing and damaging a decorative finish, then some form of access trap may be more desirable. Figure 9.9 gives some examples on how simple access traps may be provided which will allow access to regulating or stop vales.

The access traps themselves, may be opened via a hinge or similar, and held closed by the use of magnetic catches.

Other materials may include a plastic cap to fit into an access hole.

Note: Stop or regulating values should always have a means of access

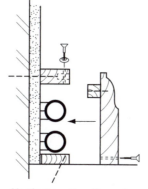

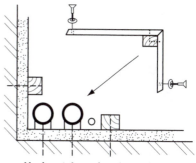

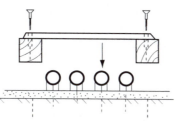

**Vertical section through a
detachable skirting board**

**Horizontal section through a corner
casing with detachable facing**

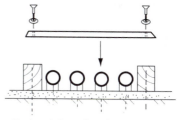

**Horizontal section through
a detachable wall casing**

**Horizontal section through a wall
casing with detachable facing**

Fig. 9.8 *Casing in pipes without a pipe backing board*

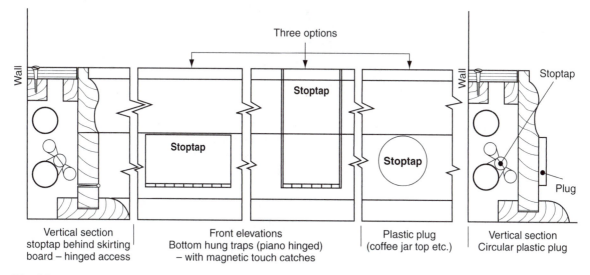

Three options

Stoptap

Stoptap

Stoptap

Stoptap

Wall

Wall

Stoptap

Plug

Vertical section
stoptap behind skirting
board – hinged access

Front elevations
Bottom hung traps (piano hinged)
– with magnetic touch catches

Plastic plug
(coffee jar top etc.)

Vertical section
Circular plastic plug

Fig. 9.9 *Accessing a stoptap behind skirting board*

9.1.7 Other casings

Window and door linings (see Chapters 3 and 4) may also be referred to as 'casings' just as double-hung sash windows, which use weights housed within a 'cased' or 'boxed frame', are sometimes called 'cased window frames'.

9.2 Wall panelling

Broadly speaking panelling may, like, many items of temporary work, or work which replaces or otherwise serves as a permanent substitute for the original surface, may be regarded as "falsework."

Figure 9.10 shows how falsework may be used as a 'cover-up'. For example, the ceiling is false and is suspended from the floor or roof above to mask pipework or heating ducts, etc. The wall panelling in this case has been fixed over the plastered or self-finished walls to mask the original surface. Purpose-made bath panels to cover the plumbing system are nearly always available as an optional extra with the bath, but there are instances where these cannot be used, for example with partly sunken baths or raised baths with a stepped frontage.

Note: *Because of its temporary nature, items of formwork (Vol. 2, Chapter 3) are also referred to as 'falsework'*

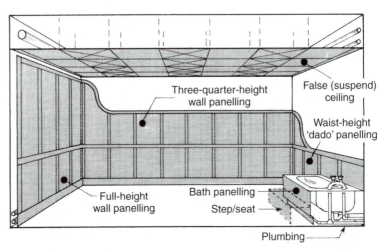

Fig. 9.10 *Examples of falsework and panelling*

Suspended ceilings are now mostly constructed from lightweight proprietary systems which use a grillwork of extruded or pressed metal sections on to or into which panels of plasterboard, fibreboard, or plastics either sit or are slotted. These systems are outside the scope of this book – if further information is required, refer to the British Gypsum handbook and other trade literature.

Wall panelling may be used to:

(a) Conceal unsightly services or uneven surfaces;
(b) Provide a decorative feature;
(c) Provide hygiene and easy clean surfaces;
(d) Provide special acoustic properties;
(e) Incorporate thermal installation;
(f) Provides a resilient and hardwearing surface.

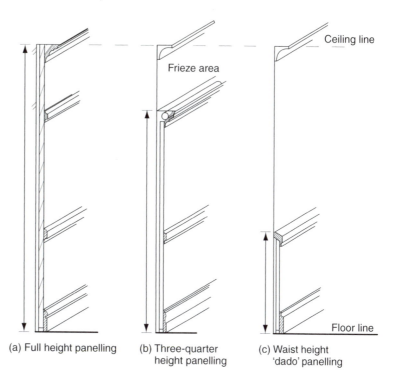

Ceiling line

Frieze area

Floor line

Fig. 9.11 *Wall panelling by height*

(a) Full height panelling

(b) Three-quarter height panelling

(c) Waist height 'dado' panelling

The panelling may also be:

(a) Full height of the room (Fig. 9.11a);
(b) Three-quarter room height – leaving the upper portion as a frieze (Fig. 9.11b);
(c) Waist height (about 1 m above the floor), better known as dado panelling (Fig. 8.1 and Fig. 9.11c).

Panelling of various heights can be further subdivided by type – in Fig. 9.12 waist height panelling is used as an example, thus:

(a) *Framed panelling (Fig. 9.12a):* a framework of timber grooved to house a series of panels;
(b) *Strip panelling (Fig. 9.12b):* made up of a series of grooved muntins to hold intermediate panels, or horizontal or vertically positioned tongued-and-grooved matchboard (see also Fig. 2.1a–d);
(c) *Sheet panelling (Fig. 9.12c):* large panels face fixed to timber grounds.

9.2.1 Timber groundwork for wall panelling

All three methods of wall panelling will require an arrangement of timber grounds on to which panel work can be attached. The use of grounds enables the minimum amount of wall fixings to be made, while allowing panel work fixings to be made at whatever intervals are required – usually determined by the thickness of the panel work.

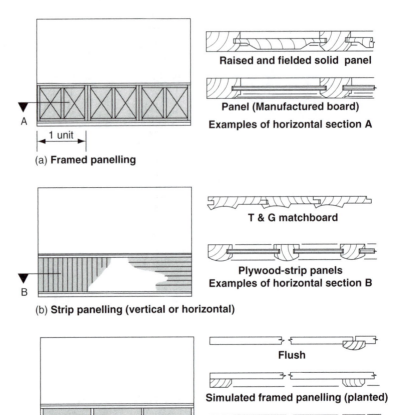

Raised and fielded solid panel

Panel (Manufactured board)

Examples of horizontal section A

(a) Framed panelling

T & G matchboard

Plywood-strip panels
Examples of horizontal section B

(b) Strip panelling (vertical or horizontal)

Flush

Simulated framed panelling (planted)

Framed
Examples of horizontal section C

(c) Sheet panelling

Fig. 9.12 *Three types of waist height (dado) panelling*

There are two main types:

- Unframed;
- Framed.

Unframed grounds (Fig. 9.13a): consist of a series of timber laths individually fixed to the wall at pre-determined centres; these may be positioned vertically or horizontally. Grounds must be fixed plumb and in line (use of a straight edge or builders line, see Volume 2, section 1.2) – great care must be taken to ensure that internal and external angles are always plumb.

Framed grounds: a prefabricated open framework of timber jointed and assembled as shown in Fig. 9.13b. Grounds should be plumbed and levelled with the aid of wood or plastics packing, then fixed to the wall via plugs or direct fixings. The use of nails or screws will depend on the plug type and base material.

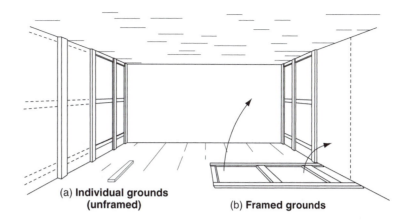

(a) **Individual grounds (unframed)**

(b) **Framed grounds**

Fig. 9.13 *Unframed (individual) and framed grounds*

The use of cartridge operated fixing tools (Volume 1, section 8.3) may be considered if the base material is suitable.

Whether individual or framed grounds are used (Fig. 9.13) will depend on the base material and the amount and type of panelling. In any case (with the exception of strip panelling), grounds should be positioned behind each vertical and horizontal joint. The number of intermediate grounds and the distance between them will depend on the permitted span of the panelling material.

Grounds for dado and three-quarter panelling can be bevelled, as shown in Fig. 9.19, to form a key for plaster. Suitable tee joints for framed grounds are shown in Fig. 9.14; corner joints should be adapted accordingly. Venting behind panels via groundwork is essential where rooms are liable to be subjected to high humidity. Vents may be sited at about skirting board height and extend to the height of the panel. They must not extend into the ceiling or floor (false or otherwise) otherwise, in the event of a fire, the venting channels could act as a flue and cause the fire to rapidly spread to other rooms.

9.2.2 Fixing panelling to timber grounds

Fixings may be made by one or a combination of the following methods. Concealed drop-on fixings (Fig. 9.15) illustrates four methods:

(a) Tongued blocks (plywood buttons) screwed on to the rear face of the panelling and dropped into edge-grooved groundwork (Fig. 9.15a);
(b) Cranked metal plates screwed on to the rear face of the panelling and again using grooved groundwork (Fig. 9.15b);
(c) Bevelled blocks cut to fit over bevelled grounds (Fig. 9.15c);

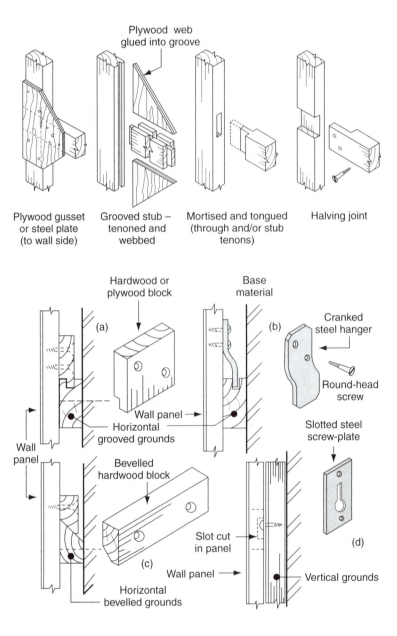

Fig. 9.14 *Framing joints for framed grounds*

Fig. 9.15 *Fixing panels using drop-on fixings*

(d) Slotted metal plates fixed to the rear of the panel work, allowing it to be dropped on to protruding screw heads set at pre-determined centres (Fig. 9.15d).

Finally, slot screwing as described in Volume 1, section 10.2.1e may also be used.

9.2.2.1 Concealed face fixings

Figure 9.16 shows several methods of concealing face fixings. The use of screws is most common, but nail heads can be concealed in a

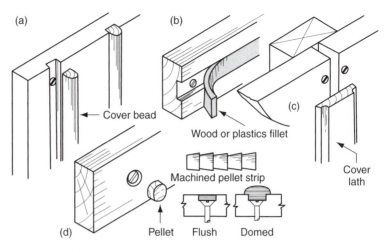

Fig. 9.16 *Concealing face fixings*

similar way. The best methods are to use a bead (Fig. 9.16a), a fillet of wood or plastics (Fig. 9.16b), or a cover lath (Fig. 9.16c). The use of counter-bored holes plugged with pellets cut across the grain of a matching species of wood is another alternative (Fig. 9.16d).

9.2.2.2 Panel adhesives and adhesive pads (Fig. 9.22)

These are most suited to lightweight panelling – manufacturers' instructions should always be consulted with regard to their use and application. If the wall surfaces are sound and plumb, timber grounds may in some cases be dispensed with.

9.2.3 Framed wall panelling

Often referred to as 'traditional' panelling, this is made up of a series of framed units consisting of a framework of grooved or rebated solid timber (usually joined together with mortises and tenons), which houses one or more solid or plywood panels. Panel shapes are often as shown for panelled doors (see Fig. 4.14).

Traditional panel work is usually made of hardwood and is prefabricated in the workshop into unit sizes suitable for safe transportation and easy access into the premises in which it is to be fixed. Examples are shown in Figs 9.12a and 9.17. Units are normally partly polished before they are fixed and are finished on site when the danger of surface damage is at a minimum.

9.2.4 Strip wall panelling (Fig. 9.12b)

This type of panelling can have the effect of making walls appear longer or taller, depending on whether the boards are laid horizontally or vertically. The direction of the groundwork will predominately be at right angles to the run of the strip panelling.

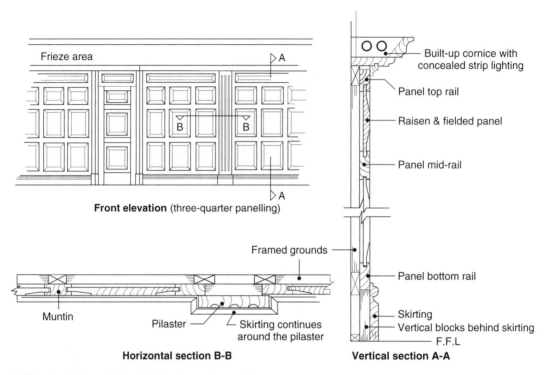

Frieze area

▷A

B   B

▽A

Front elevation (three-quarter panelling)

Framed grounds

Muntin

Pilaster —— Skirting continues
around the pilaster

Horizontal section B-B

Built-up cornice with
concealed strip lighting

Panel top rail

Raisen & fielded panel

Panel mid-rail

Panel bottom rail

Skirting
Vertical blocks behind skirting
F.F.L

Vertical section A-A

Fig. 9.17 *Traditional frieze height framed panelling*

Wood in the form of tongued-and-grooved boards or plywood is the
most common strip material, but metal and plastics profiles are also
popular. Reference should be made to 'External cladding' (Chapter
2) for the different profiles and methods of concealing the fixings.
Figure 9.18 shows two examples of secret fixing vertical or horizontal tongued-and-grooved boarding (strip panelling).

9.2.5 Sheet wall panelling (Fig. 9.12c)

This modern panelling – often referred to as 'flush' panelling – consists of manufactured boards in their natural state or faced with
another material. For example, typical base materials would include
blockboard, laminboard, MDF, plywood, particle boards (chipboard), and different fibreboards such as softboard (insulation
board), medium boards, and hardboards. Finishes such as polished
wood veneers, stainless steel, brushed aluminium, plastics, plastics
laminates, and various woven fabrics are very popular.

Figure 9.19 shows vertical sections through three heights of sheet
panelling. The full-height panelling (Fig. 9.19a) is left short at the top
to allow for ventilation if needed and is covered by a scribed fillet or
cornice. If a suspended ceiling is to be incorporated, allowance can
also be made at the head. The plinth is recessed.

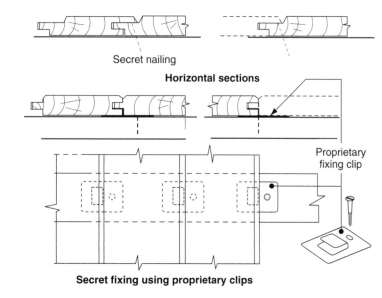

Horizontal sections

Secret nailing

Proprietary fixing clip

Fig. 9.18 *Secret fixing tongued-and-grooved boarding*

Secret fixing using proprietary clips

Part vertical elevation

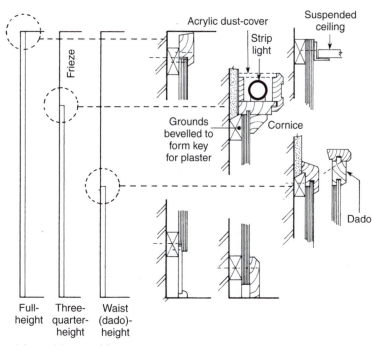

Acrylic dust-cover

Suspended ceiling

Strip light

Frieze

Grounds bevelled to form key for plaster

Cornice

Dado

Fig. 9.19 *Vertical sections through sheet wall panelling at different heights*

Full-height   Three-quarter-height   Waist (dado)-height

The three-quarter panelling (Fig. 9.19b), shows how concealed strip lighting can easily be set into (behind) the cornice to illuminate the frieze and ceiling. A skirting board has been used to cover the gap left under the panelling, which can be used to ventilate the back of the panelling.

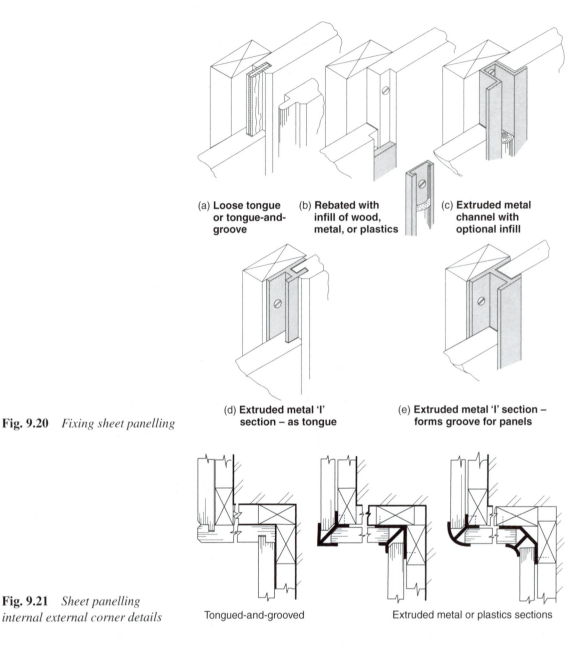

(a) **Loose tongue or tongue-and-groove**

(b) **Rebated with infill of wood, metal, or plastics**

(c) **Extruded metal channel with optional infill**

(d) **Extruded metal 'I' section – as tongue**

(e) **Extruded metal 'I' section – forms groove for panels**

Fig. 9.20   *Fixing sheet panelling*

Fig. 9.21   *Sheet panelling internal external corner details*

Tongued-and-grooved   Extruded metal or plastics sections

Figure 9.19c shows how a dado rail trims the top edge of the dado panelling.

Different methods of joining panels vertically are shown in Fig. 9.20. Figure 9.21 shows how internal and external angles can be treated.

Thin sheets of suitably faced plywood or fibreboard may be used as panelling for reasons of lightness or economy. Figure 9.22 shows some

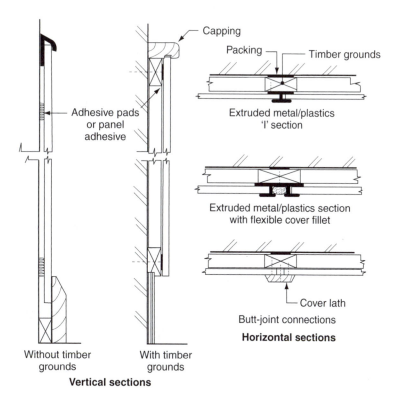

Capping

Packing — Timber grounds

Extruded metal/plastics
'I' section

Extruded metal/plastics section
with flexible cover fillet

Cover lath

Butt-joint connections

Horizontal sections

Adhesive pads
or panel
adhesive

Fig. 9.22 *Fixing thin sheet panelling*

Without timber
grounds

With timber
grounds

Vertical sections

possible methods of fixing them to the base material and of making sheet-to-sheet connections with or without the use of timber grounds.

9.3 Fire performance (rate of flame spread)

If a fire breaks out within the confines of a room with walls and ceiling clad with a combustible material, it is obvious that the risk of flames spreading over the surfaces will be much greater than if they had been covered with a non-combustible material like plaster. Therefore, because many of the materials used for wall panelling are to some extent combustible, limitations are in some cases put on their use, unless they are suitably treated with a fire retardant to satisfy current Building Regulations.

Rate of flame spread and its classification are discussed at Chapter 2, section 2.2.6, Fire performance.

9.4 Bath panelling

The type of panel will depend on where the bath is sited in the room, for example:

(a) With only one long side exposed;
(b) With one long side and one end exposed;
(c) With two long sides and one end exposed.

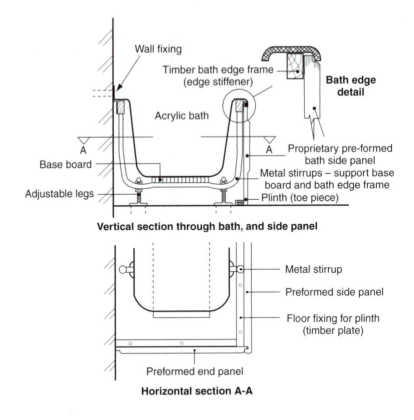

Fig. 9.23 *Fixing pre-formed bath panel*

We are going to consider two types of panelling:

- pre-formed panels made from various materials;
- panels built-up *in-situ*.

9.4.1.1 Pre-formed panels (Fig. 9.23)

These usually only require a small-sectioned timber plate attached to the floor as a restraint and to provide a means of fixing the bottom edge of the panel or toe piece. The top edge is held in place by a groove formed by a pre-fixed timber or metal section stiffened to the underside of the rolled edge of the bath. The arrangement and fixing of the stiffener can vary between makes of bath.

9.4.1.2 Built-up bath panels (Fig. 9.24)

These will require a supporting timber back framework constructed with simple framing joints (Volume 1, section 10.3) secured to the under edge of the bath stiffener and the floor. It is important that these frames are not over-tightened between these points otherwise

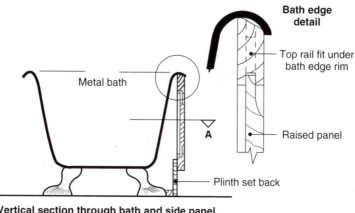

Bath edge detail

Metal bath

Top rail fit under bath edge rim

A

Raised panel

Plinth set back

Vertical section through bath and side panel

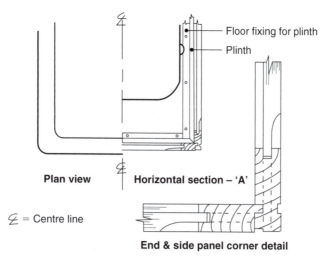

Floor fixing for plinth

Plinth

Plan view

Horizontal section – 'A'

$\mathcal{L}$ = Centre line

End & side panel corner detail

Fig. 9.24 In-situ *built-up bath panel*

there will be a tendency to lift the front edge of the bath and possibly cause it to distort.

The panelling material can be attached to this subframework. The panel material may be in the form of a framed, strip, or sheet panel (see sections 9.2.3–5), with various decorative finishes.

References

BS 1282: 1999, Wood Preservatives. Guidance on choice use and application.

BS 4261: 1999, Wood Preservation, Vocabulary.

BS 5589: 1989, Code of practice for wood preservation.

BS 5707: 1997, Specification for preparations of wood preservatives in organic solvents.

BS 476–7: 1997, Fire tests on building materials (surface spread of flame).

BS EN 634–2: 1997, Cement-bonded particle board, specification.

BS EN 636: 2003, Plywood specifications.

BS 5973: 1993, Code of practice for access and working scaffolds and special scaffold structures in steel.

BS 1139–3: 1994, Specification for prefabricated access and working towers.

BS 1139–4: 1982, Specification for prefabricated steel splitheads and trestles.

BS 1129: 1990, Timber ladders, steps, trestles, and lightweight stagings.

BS 2037: 1994, Aluminium ladders, steps, and trestles for the building and civil engineering industries.

BS 2482: 1981, Specification for timber scaffold boards. Building Regulation Approved Document B, Fire safety: 2000.

BS 5973: 1993, Section 9 – tying to buildings.

HSE Information Sheet General Access, Scaffolds and Ladders Information Sheet 49.

Building Regulation Approved Document B, Five Spread: 2000.

10 Joinery fitments and purpose-made joinery

Examples of how some domestic fitments may appear are shown in Fig. 10.1. It is usual for each group of fitments to be named according to their location, for example:

(a) kitchen fitments;
(b) bathroom fitments;
(c) bedroom fitments.

Fitments are, of course, found in other areas of the home, such as a lounge or a dining room but, because their use is often multi-purpose, they are less specifically named. Such an item could consist of a fitted base or wall unit – even a shelved room-divider – which might at sometime be expected to house among other things a television, video, and/or hi-fi systems.

Joiners can also be expected to fit out offices, workshops, and warehouses of industrial premises. Commercial premises such as shops and banks, etc. usually employ specialist shop fitters, kitchen fitters or joinery firms which specialize in purpose-made joinery products.

Broadly speaking, joinery fitments can be divided into six groups:

- *Shelving:* these are fitted between alcoves or supported on uprights;
- *Free-standing units:* these stand off the floor without added support;
- *Hung units:* these rely upon a wall and/or ceiling for support;
- *Fitted units:* these are 'free-standing' or 'hung' units which – because of possible instability when in use or the need to fit close to a wall, floor, ceiling, or another unit – require fixing to whatever they abut;
- *Built-in fitments:* these rely fully or in part upon their attachment to a floor, wall, or ceiling, or a combination thereof, for the formation of the structural framework and/or the stability of the fitment (Fig. 10.9).

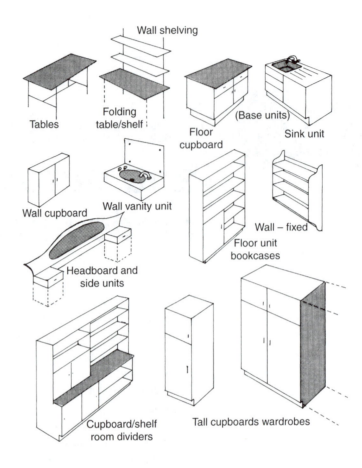

Fig. 10.1 *Examples of some joinery fitments*

10.1 Shelving

Shelves are expected to support the differing weight-to-volume ratios of a vast variety of items and materials. It is therefore essential that, before shelves are constructed or assembled, consideration is given to both their function and location as these are the essential factors that must be integrated within the overall design.

Figure 10.2 a shows the dangers of:

(i) using shelving of inadequate thickness or strength to withstand the necessary imposed loads; or
(ii) expecting shelving to span an unrealistic distance.

Figure 10.2b and 10.2c can be regarded as suitable remedial alternatives, i.e. providing the necessary intermediate support and/or using stronger material.

Shelves are made up of either 'solid' or slatted material, examples of which are shown in Fig. 10.3 (see also Volume 1, section 10.2.). Over the years, various methods of supporting shelves have evolved which the joiner can adapt to suit their specific requirements. A few examples of

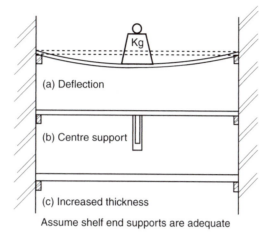

(a) Deflection

(b) Centre support

(c) Increased thickness

Assume shelf end supports are adequate

Fig. 10.2 *Shelf support*

these methods will be found in sections 10.1.2 and 10.1.3. (You should also consult Volume 1, Chapter 12 with regards to fixing devices.)

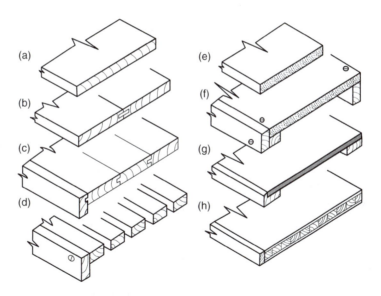

Fig. 10.3 *Shelfboards and their construction*

Note: *With all types of manufactured boards, appropriate PPE must be worn as well as consulting the COSHH Regulations when cutting or shaping*

Key for Fig. 10.3:

(a) single board;
(b) double board with loose tongue;
(c) tongue-and-grooved with front stiffener;
(d) slatted with front stiffener;
(e) particle board or MDF – self-finish or with veneered wood or plastics;
(f) particle board or MDF with front and back stiffeners;

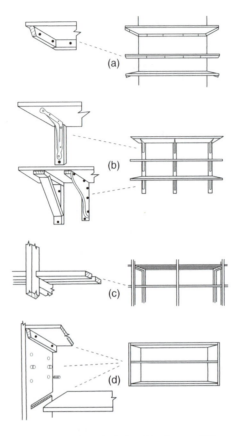

Fig. 10.4 *Traditional methods of shelf support and assembly*

(g) plywood or MDF with front stiffener and back support;
(h) blockboard or laminboard with slipped front edge.

10.1.2 Traditional shelving

As can be seen from Fig. 10.4, shelves can be supported by:

(a) bearers fixed to wall;
(b) brackets – metal, or purpose-made from timber and/or plywood;
(c) timber – framed uprights (standards);
(d) solid uprights (enclosed units), where provision for shelf adjustment can be made.

The methods in Fig. 10.4a and 10.4b rely on walls for their support, whereas those in Fig. 10.4c and 10.4d may be free-standing units.

10.1.3 Proprietary systems and shelving aids

Commercially produced forms of shelf support may follow a similar pattern to those in Fig. 10.4b and 10.4d, but provide for greater flexibility, in both their construction and the means of shelf adjustment.

Figure 10.5a shows how metal brackets of various styles and sizes can be positioned to any height by hooking them onto an upright metal channel which has been secured to a wall. Figure 10.5b shows how a series of holes bored in a solid upright enables metal or plastic shelf-supporting studs and their sockets to be positioned to suit shelf

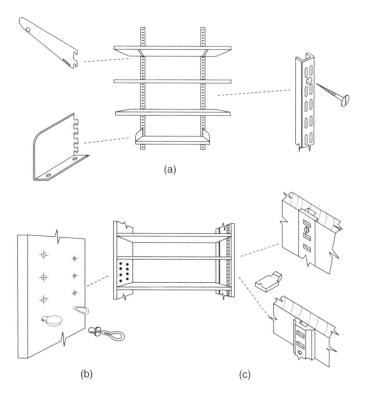

Fig. 10.5 *Proprietary methods of shelf support*

(a)

(b)   (c)

requirements. Alternatively, a metal strip can be housed or surface fixed with adjustable pegs as shown in Fig. 10.5c.

10.2 Designing fitments

The following considerations should always be taken into account before any work begins. First, it is necessary to ensure that fitment heights suit the user, i.e. an adult, a child, or in some cases a disabled person. All work surfaces, cupboards, and shelving in regular use should be easily accessible without the need for excessive bending or stretching. Wall-cupboard doors should be made to slide, unless they can be hinged opened without the risk of a person accidentally walking into an open door. Tall narrow units should be fastened back to a wall or up to a ceiling to prevent them toppling over – such units are particularly at risk when all the hinged doors are open. (There has always been this danger with some free standing wardrobes – especially if they have heavy mirrored doors.)

Secondly, the width and length of fitments are important, especially when units are designed to house sinks, vanity basins, or gas or electrical appliances. If base units are to be backed, then an allowance of not less than 50 mm should be left between the back and the wall for pipe-work and skirting board, etc. Figures 10.6 and 10.7a and b give some guidance as to various accepted unit sizes.

Finally, the type and method of construction should be considered, as below.

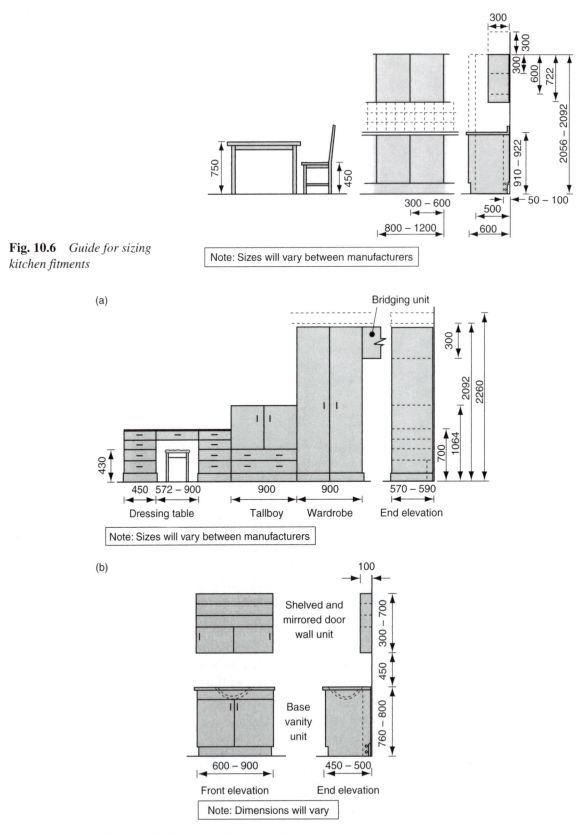

Fig. 10.6 *Guide for sizing kitchen fitments*

Note: Sizes will vary between manufacturers

(a)

Bridging unit

Dressing table

Tallboy   Wardrobe   End elevation

Note: Sizes will vary between manufacturers

(b)

Shelved and mirrored door wall unit

Base vanity unit

Front elevation   End elevation

Note: Dimensions will vary

Fig. 10.7 *Guide for sizing bedroom and bathroom fitments*

10.2.1 Carcase construction

'Carcase work' generally refers to the main framework of the whole unit but – because they form part of the main framework with some methods of construction – the top and plinth sections will also be dealt with under this heading.

10.2.1.1 Worktables

As shown in Fig. 10.8, there are four main considerations:

(i) the joint between the rails and the table legs;
(ii) whether to include tie rails;
(iii) whether to include a drawer or drawers;
(iv) how to attach the table top.

By mortising the rails into the legs (Fig. 10.8a) a strong joint can be made. Its strength is determined by the rail depth (tenon depth) and the leg section (tenon length) – effective length is increased by mitring the ends of each tenon as shown. The use of a top mid-rail (Fig. 10.8b) dovetail-housed into the front and back rails will depend on the table length.

A much simpler method of corner jointing is shown in Fig. 10.8c, which uses a Crompton table-plate to hold the rails in position. The legs are bolted to the plate via a steel washer and a wing-nut – studs having previously been screwed into the corner of each leg. The main advantage of this method is that the legs can be easily dismantled at any time simply by unscrewing the wing-nuts.

Tie rails prevent table legs from spreading. Legs placed upon smooth surfaces like a vinyl-covered floor will stress the joints between the rails and legs, particularly if the legs are splayed. Therefore, provided they do not impede the users' leg room, ties will prove an asset.

Drawers can be included within the design, but can add considerably to the overall cost of the table. Figure 10.8d shows how the construction can be modified to include table-top drawers. The way in which the table-top is attached will depend on what material the top is made of and whether there is any likelihood of moisture movement of the top after fixing. Manufactured boards (Fig. 10.8e) present little difficulty, therefore any of the methods shown in Fig. 10.8g could be adopted. On the other hand, if a solid timber top (suitably jointed) was required (Fig. 10.8f), then rigid fixings using pockets and blocks should be avoided.

Round-head screws through steel washers should be used where steel movement plates meet the underside of the table-top. Turn-buttons are screwed to the underside of the table-top, then – when the top is in position – the buttons are turned until their tongues fit into the grooves or slots cut into the backs of the top rails. The screws can then be tightened.

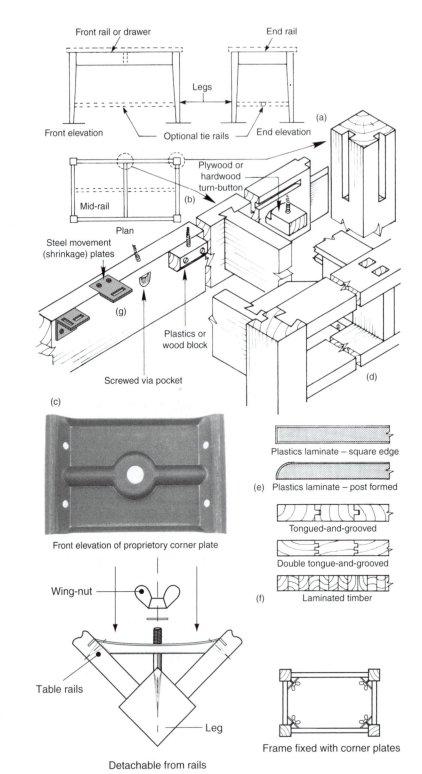

Front rail or drawer

End rail

Legs

Front elevation   Optional tie rails   End elevation   (a)

Mid-rail

Plan   (b)

Plywood or hardwood turn-button

Steel movement (shrinkage) plates

Screwed via pocket   (g)

Plastics or wood block   (d)

(c)

Front elevation of proprietory corner plate

Wing-nut

Table rails

Leg

Detachable from rails

Plan view of leg attached prior to securing with wing-nut

Plastics laminate – square edge

(e)   Plastics laminate – post formed

Tongued-and-grooved

Double tongue-and-grooved

(f)   Laminated timber

Frame fixed with corner plates

Fig. 10.8 *General-purpose work tables*

10.2.1.2 Purpose made base units (free-standing or fitted)

Figure 10.9 shows four methods of constructing these units:

Boxed or rail-and-slab construction (Fig. 10.9a) is the method favoured by most unit manufacturers. Base, ends, and intermediate upright panels and shelving are all cut from sheet material – usually 16 mm thick plastics-faced particle board. Modular sizes usually reflect standard metric sheet sizes. Unit backs, which hold the unit square, are usually made from pre-finished hardboard, particle board or MDF.

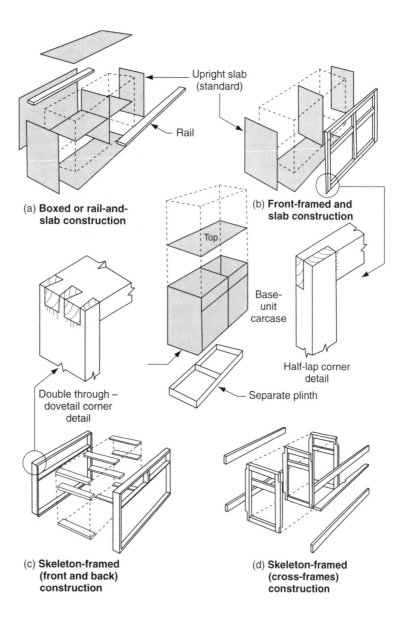

Fig. 10.9 *Methods of constructing base units for fitted or freestanding cupboards*

(a) **Boxed or rail-and-slab construction**

(b) **Front-framed and slab construction**

(c) **Skeleton-framed (front and back) construction**

(d) **Skeleton-framed (cross-frames) construction**

Upright slab (standard)

Rail

Top

Base-unit carcase

Half-lap corner detail

Separate plinth

Double through – dovetail corner detail

Most manufacturers generally prefer to use proprietary knock-down fittings (see Fig. 10.12b), which enable the whole unit to be packed flat. All that is then required for on-site assembly is the ability to understand the manufacturer's drawings and usually the use of a screwdriver or a special tool supplied. Originally these knock-down fittings were available only to the manufacturers of assembly-line fitments, but similar fittings are now freely available from most hardware suppliers. The plinth may be separate or built into the carcase.

10.2.1.3 *Front-framed and slab construction (Fig. 10.9b)*

Again, the ends and intermediate upright panels are of sheet material. A back is optional. The front is framed-up from small timber sections (e.g. 25 mm × 50 mm) jointed by mortise-and-tenon and/or halving joints. The plinth may be separate or built into the carcase.

10.2.1.4 *Skeleton-framed (front and back) construction (Fig. 10.9c)*

Front and back frames are joined to cross-rails which serve as shelf bearers and/or drawer runners. The plinth is usually separate.

10.2.1.5 *Skeleton-framed (cross-frames) construction (Fig. 10.9d)*

Framed uprights serve as shelf-bearers or supports for drawer runners. They are positioned to suit door or drawer lengths and are notched to receive front and back runners. The plinth may be separate, but is usually formed by notching back to take a toe piece.

Figure 10.10 shows a unit made of manufactured board to include housing joints to receive the fixed shelf and baseboard (traditionally known as the *potboard*) which are glued, pinned and later filled with a suitable filler.

To support the backboard of the unit, a groove is machined at the back of each upright (or standard).

A front and back cross-rail may be included that has been dovetailed into the top of the uprights for extra strength. Screws may be used through these rails to assist with the fixing of the top rather than using fixing blocks (Figs 10.8 and 10.10d).

Manufactured board often requires edging, as shown in Fig. 10.11 especially in the case of particle board or blockboard; for this reason edgings of either natural wood (Fig. 10.10b) fitted into grooves along all exposed edges, or plastic laminated strip (Fig. 10.10c) which are applied by various means using heat. All these edged finishes should fully match the unit.

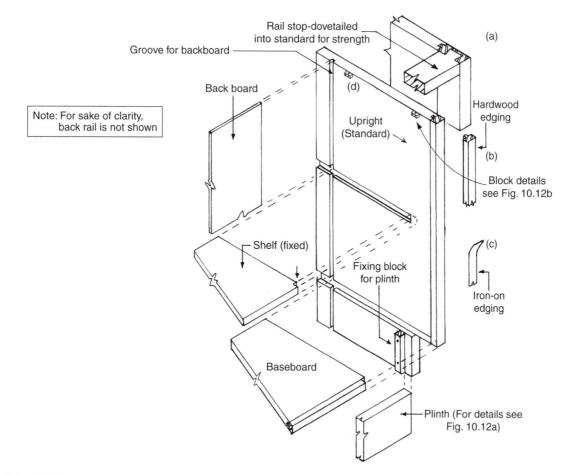

Rail stop-dovetailed into standard for strength (a)

Groove for backboard

(d)

Back board

Note: For sake of clarity, back rail is not shown

Hardwood edging

Upright (Standard)

(b)

Block details see Fig. 10.12b

Shelf (fixed)

(c)

Fixing block for plinth

Iron-on edging

Baseboard

Plinth (For details see Fig. 10.12a)

Fig. 10.10 *Cupboard unit with traditional jointing methods*

Figure 10.11 shows several arrangements for edges of panels.

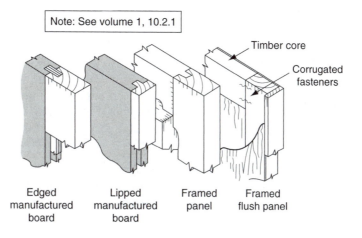

Note: See volume 1, 10.2.1

Timber core

Corrugated fasteners

Fig. 10.11 *Fabricating end panels and divisions (standards or uprights)*

Edged manufactured board

Lipped manufactured board

Framed panel

Framed flush panel

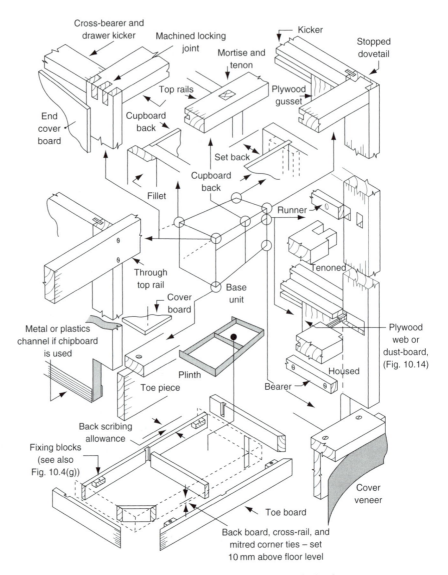

Fig. 10.12a   *Alternative joints for carcase and plinth*

10.2.2 Joining members and components

Figure 10.12a shows how traditional joints between frames, slabs, and rails can be made. Alternatively, proprietary knock-down connectors could be used, but many of these can be unsightly when viewed from inside the unit (Fig. 10.12b).

Note: *See also Volume 1, section 10.3, Framing – angle joints*

10.2.3 Built-in fitments

Figure 10.13 shows a tall fitted corner cupboard suitable as a single wardrobe or airing cupboard. The cupboard consists of a prefabricated front frame with pre-hung doors and one handed side panel (to

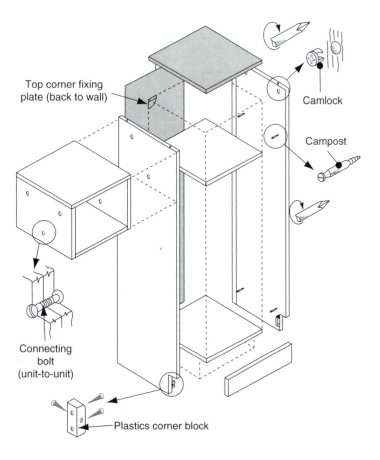

Top corner fixing
plate (back to wall)

Camlock

Campost

Connecting
bolt
(unit-to-unit)

Plastics corner block

Fig. 10.12b   *Joints for slab construction using proprietary fittings*

suit either left- or right-hand walls). Existing walls serve as a back and the other side. The drawing shows two alternative types of door, and the front frame can differ in the way that the bottom rail is attached to the toe board when forming a plinth.

Both the front frame (without doors) and the side frame are plumbed, then scribed to the walls. Positions of wall bearers (shelf positions) and uprights are levelled, marked, and (where necessary) plugged to the wall to receive shelves and the edge of the side panel. The front frame is then fixed to the front edges of the shelves and the side panel. The cornice or trim is then fixed between the cupboard and the ceiling, then the doors are re-hung and the handles and catches are fitted.

10.3 Wood drawers

It has always been regarded as a major achievement when an apprentice joiner can successfully make and fit, by traditional means, a drawer into a pre-formed drawer opening, because the operation as a whole requires a considerable amount of skill.

10.3.1 Drawer openings

These should match the type of drawer. For example, Fig. 10.14 shows a cut-away view of a traditional drawer opening designed to

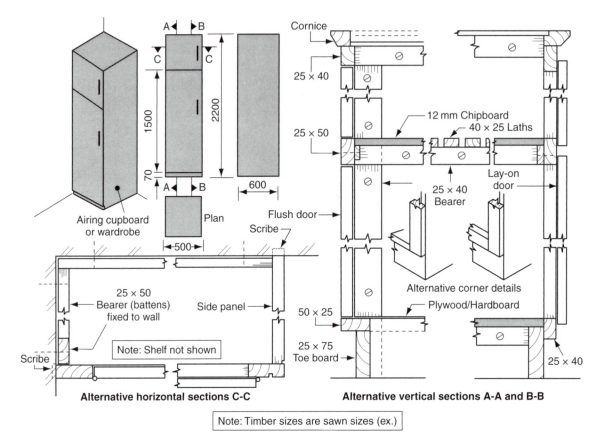

Airing cupboard or wardrobe

Plan

Flush door

Scribe

Alternative horizontal sections C-C

25 × 50
Bearer (battens)
fixed to wall

Side panel

Note: Shelf not shown

Scribe

Cornice

25 × 40

25 × 50

12 mm Chipboard
40 × 25 Laths

25 × 40
Bearer

Lay-on
door

Alternative corner details

Plywood/Hardboard

50 × 25

25 × 75
Toe board

25 × 40

Alternative vertical sections A-A and B-B

Note: Timber sizes are sawn sizes (ex.)

Fig. 10.13 *Fitted corner cupboard*

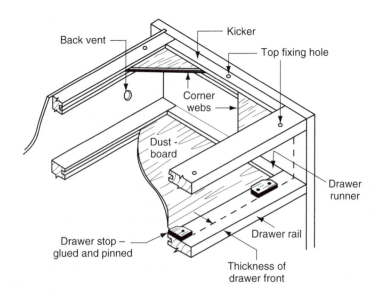

Back vent

Kicker

Top fixing hole

Corner
webs

Dust
board

Drawer
runner

Drawer stop –
glued and pinned

Drawer rail

Thickness of
drawer front

Fig. 10.14 *Traditional drawer openings*

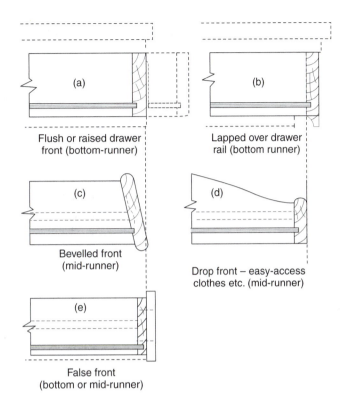

(a)

Flush or raised drawer
front (bottom-runner)

(b)

Lapped over drawer
rail (bottom runner)

(c)

Bevelled front
(mid-runner)

(d)

Drop front – easy-access
clothes etc. (mid-runner)

(e)

False front
(bottom or mid-runner)

Fig. 10.15 *Types of drawer front*

house the whole of the drawer front (Fig. 10.15a). If the drawer recess consists of solid sides, top, and dust-board bottom, the action of the close-fitting drawer compressing the air behind it could make drawer closing awkward, unless the drawer back is lower than the sides or a vent hole is cut into the back of the carcase.

In all cases, the run of the drawer is checked by gluing and tacking two hardwood or plywood drawer stops near the ends of each drawer rail as shown.

Figure 10.15 shows how different fronts can be used. Figure 10.15a may be set in line with or raised outside the carcase – it slides over a bottom runner. Figure 10.15b is similar, but the drawer rail is set back the thickness of the drawer front – in this way the drawer rails within a nest of drawers (drawers set one upon another) may be omitted or concealed. With single drawers, the overhang below the drawer sides can be cut away to form a drawer pull. The drawer front in Fig. 10.15c is sloping and drops below the sides; therefore the drawer rail can again be omitted or concealed – the runners in this case are inset midway into the drawer sides. Figure 10.15d has a drop front to allow easy access to clothing, etc. and is often found within a wardrobe.

10.3.1.1 Drawer runners (Fig. 10.16)

Traditionally, drawer runners were built into the main frame carcase, and the bottom of the drawer sides slid over them, wood on wood (Fig. 10.16a). Candlewax was used as a lubricant. Drawers in constant use eventually became worn, with the result that the drawer front became displaced.

Hardwood inset mid-runners (Fig. 10.16b) use a groove cut into the drawer sides. They are also subject to wear.

Modern runners are available in many different forms. Some are made from plastics (Fig. 10.16c) or of fibre-reinforced plastics (Fig. 10.16d) – both give easy drawer movement. Others are made of metal, with intervening steel or plastics ball-races to give the effort-free motion we associate with the drawers of modern filing cabinets. Hardwood top-hung

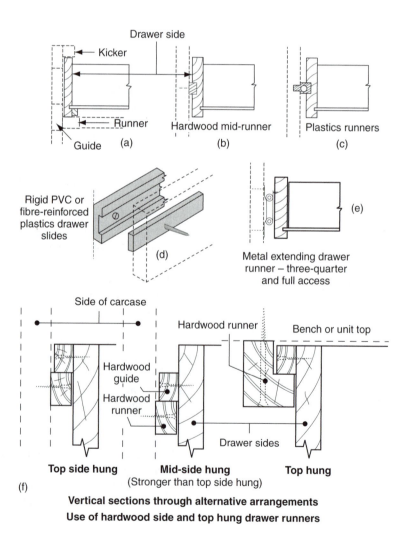

Fig. 10.16 *Drawer runners*

track (Fig. 10.16f) is ideal for adding a drawer under a worktop or table, but is only intended for use with shallow drawers with light contents.

10.3.1.2 Drawer joints (Fig. 10.17)

Drawer sides to a non-overhanging front will normally be dovetailed (worked either by hand or by machine) and stopped (lapped dovetails) if the face of the drawer is to show, otherwise through dovetails can be used. Overhanging drawer fronts can be trenched to receive a dovetailed housing, or planted on as a false front to a through-dovetailed framework.

The drawer back can be joined to the sides by using either through dovetails or stopped housings. The drawer bottom fits into grooves cut into the drawer sides and front (or into add-on beads as shown in Fig. 10.18(5a)) and is then nailed on to the bottom edge of the drawer back. Glue blocks are stuck between the sides and the bottom.

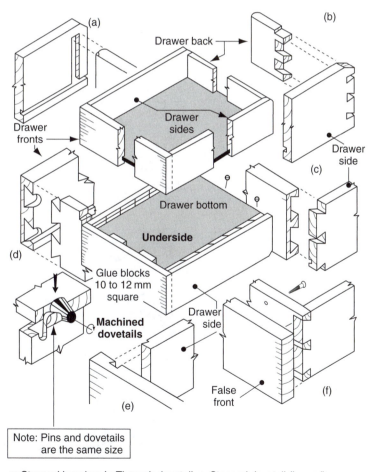

a: Stopped housing; b: Through dovetail; c: Stopped dovetail (lapped);
d: Machined dovetail; e: Dovetail housing; f: Through dovetails and false front.

Fig. 10.17 *Drawer joints*

Modern drawers come nowadays, together with bedroom units, kitchens units etc., usually ready assembled. Thus reducing the installation time by qualified experienced fitters or the competent home owner.

The drawers are often made of MDF or particle board with a timber or plastics laminated veneer with ball-bearing runners and cushion closers.

10.3.1.3 Drawer construction (traditional)

The following stages refer to those shown in Fig. 10.18.

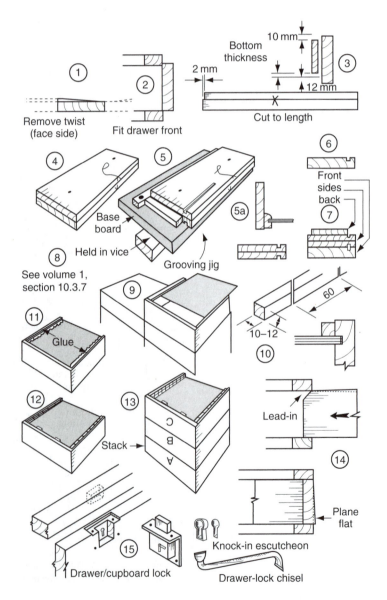

Fig. 10.18 *Stage-by-stage traditional drawer construction*

1. Ensure that the inside face of each drawer front is not twisted – if it is, remove the twist by planing.
2. Fit and label each drawer front one at a time into its appropriate opening, for example 'A', 'B', 'C', and so on. Stack to one side.
3. Cut the drawer backs shorter than the fronts by about 1.5 to 2 mm. Label to suit. Mark the face side and edge. Round the opposite edge. Stack to one side.
4. Cut the drawer sides to length. Mark the face side and edge. Tack together all matching pairs with panel pins, face sides outermost. Label then stack them to one side.
5. Groove all sides (while tacked together) to receive the drawer bottoms – use a grooving jig. Alternatively (5a), use a grooved bead.
6. Groove the drawer fronts like the sides – use the same jig.
7. Match the drawer sets together – one front, two sides (still tacked together), and one back.
8. Mark, cut, and fit all dovetail joints as described in Volume 1, section 10.3.7. Stopped housings can be used between the backs and sides (Fig. 10.17).
9. Glue (PVA) all joints – tack or cramp as necessary. For each drawer, square the drawer framework over the base carcase (drawer housing); insert the drawer bottom from the back and nail it down on to the drawer back.
10. For each drawer, cut up two drawer-depth lengths of 10 to 12 mm square section into glue blocks 60 mm long, using a mitre block. By cutting the lengths up in this way it ensures that, no matter what shape they and the bottom are, their fit into the corner is assured.
11. For each drawer, apply PVA glue along the outside joint between the drawer sides and bottom. Push the glue blocks into the corner along the length of each drawer side.
12. Add to each drawer two square-ended blocks between the underside of the drawer bottom and the drawer front.
13. Stack the drawers one on top of the other until the glue hardens.
14. Dress all the joints with a sharp smoothing plane. Plane a bevelled lead-in off the back top edge of each side. Reduce the sides to fit under the kickers. If required, dress the drawer front flush with the drawer opening. Complete one drawer at a time.
15. Fix drawer handles or pulls. If drawer locks are fitted, cut a mortise hole (use a drawer-lock chisel) into each lock rail to receive the bolts.
16. Wax (with candlewax) the drawer runners, kickers, and sides to reduce friction and wear.

> Note: *If drawer locks are to be fitted, their housings into the fronts should be cut at this stage (8).*

> Note: *Glue blocks have the effect of stiffening the drawer bottoms against 'drumming' (rattling).*

10.4 Drawers made from plastics

Once very popular, plastics drawer sides, and possibly fronts, were in the form of hollow-box extruded sections – profile examples are shown in Fig. 10.19a. Provision is made for inset drawer runners and the drawer bottom.

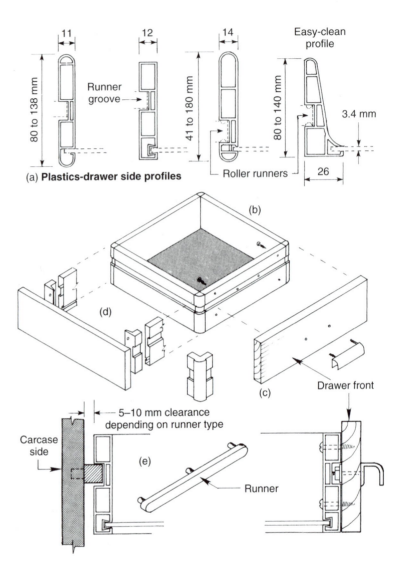

Fig. 10.19 *Drawers made from plastics*

10.4.1 Assembly (Fig. 10.19b)

The sides are cut to length and jointed with purpose-made corner blocks which fit into the hollow-box section. The joint can be made by:

(a) glueing,
(b) using non-retractable barbed metal shores;
(c) using retractable click catches which fit into pre-mortised holes cut into the drawer section. (These drawers are made to order by the manufacturer.)

Drawer fronts (other than plastics) can be attached to the profile by screws (Fig. 10.19c) or by omitting the front profile and using special end connectors as shown in Fig. 10.19d – in which case the front will have to have provision made to receive the drawer bottom.

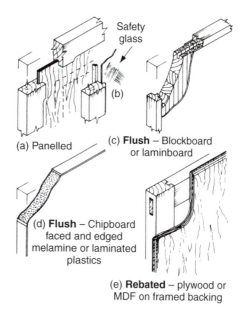

Note: All panelled doors and flush doors may now be processed from MDF

(a) Panelled

(b)

(c) **Flush** – Blockboard or laminboard

(d) **Flush** – Chipboard faced and edged melamine or laminated plastics

(e) **Rebated** – plywood or MDF on framed backing

Fig. 10.20 *Cupboard door materials*

10.4.2 Drawer runners

These range from simple plastics strips with bull-nose ends, screwed or dowelled to the sides of the drawer housing (Fig. 10.19e), to those with effortless action as a result of using rollers, ball-races and self-close mechanism.

10.5 Cupboard doors

Doors may be inset, flush, or protruding in relation to the cupboard carcase – much will depend on the door material and the front design of the cupboard. Hinges are available to satisfy most door positions. Doors can be hung on hinges or made to slide. Hung doors give quick, full, and easy access into a cupboard, but when doors are open or being opened they can be a hazard as mentioned earlier. Sliding doors, on the other hand, can restrict cupboard access, since usually only half or one-third of the front area can be opened at any one time.

Cupboard doors (Fig. 10.20) may be made by the following methods:

- *Panelled doors (Fig. 10.20a)* – a frame work of timber grooved on the inside edges to receive a panel of wood, manufactured board or safety glass (Fig. 10.20b) which must comply with current safety legislation or regulations which relate to safety applications. BS 6262-4: 1994.
- *Flush wood doors (Fig. 10.20c)* – constructed of sheet material, i.e. MDF, laminboard or blockboard, face one or both sides with wood veneer with lipped edges all round.
- *Flush particle board doors (Fig. 10.20d)* – made from chipboard that has been faced and edged with melamine or laminated plastics.
- *Rebated doors (Fig. 10.20e)* – timber framework faced with manufactured board, i.e. plywood or MDF.

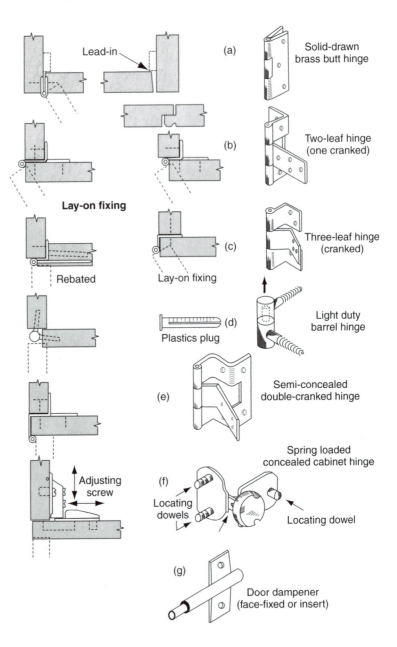

Fig. 10.21 *Hanging cupboard doors*

10.5.1 Hanging cupboard doors (Fig. 10.21)

Doors set inside their frame require butt hinges (Fig. 10.21a). For neatness, uncranked butts can be set with only half of the hinge knuckle showing – notice the lead-in on the closing edge of the door.

The two-leaf cranked hinge (Fig. 10.21b) allows a door to fit partly or fully (with a 'lay-on' fitting) over the door frame. Open doors could interfere with adjacent fitments.

Three-leaf cranked hinges (Fig. 10.21c) can used with a rebated and a lay-on fixing.

Barrel hinges (Fig. 10.21d) can be used on lift-off doors. They are fixed by screwing the metal dowel either directly or via a plastics plug into the edge of the door and the door frame.

Semi-concealed double-cranked hinges (Fig. 10.21e) allow similar cupboard units to butt one against the other – outside door edges do not interfere with adjacent closed doors.

Self-latching concealed hinges (Fig. 10.21f) do not interfere with adjacent doors. They have a built-in adjusting mechanism and are usually self-closing over the last part of the closing arc.

The Soft-close door dampener mechanism shown at Fig. 10.21g is fixed to the closing edge of the unit by two screws. Used with the spring-loaded concealed cabinet hinge it prevents the cupboard door from "*slamming*" shut, by the protruding piston that gently retracts back into its cyclinder – different types are available.

Cupboard catches – different types are available as shown in Fig. 10.22, to suit different type and sizes of door.

10.5.1.1 Fitment furniture (Fig. 10.23)

When door or drawer edge grips are not used, handles or pulls should be positioned so that the door or drawer opens or slides with ease and should be sited at a convenient position to avoid over-reaching.

Bar handles, knobs, and pulls are attached with bolts or screws. Flush pulls – associated with sliding doors – are recessed into the face of the door. Door or drawer edge grips are extruded from aluminium or plastics. Attachment is usually via a barbed tongue, which is gently driven into the grooved edge of the door or drawer. The grooves should be stopped short, and the tongues cut back so that they do not show on the door edge or drawer end.

10.5.2 Sliding doors

As shown in Fig. 10.24, arrangement of sliding doors will depend on the number of doors and the overall opening size.

Thin sliding-door panels (Fig. 10.25) – whether glass, plywood, or hardboard – use either grooves cut into their surrounding framework (plastics inserts provide a smoother action) or purpose-made channels of plastics or fibre-reinforced plastics. Whichever method is used, the top grooves will need to be deeper than the bottom grooves, to enable doors to be positioned and lifted out for cleaning purposes, etc.

Thicker and heavier doors can be run off a bottom- or top-track system.

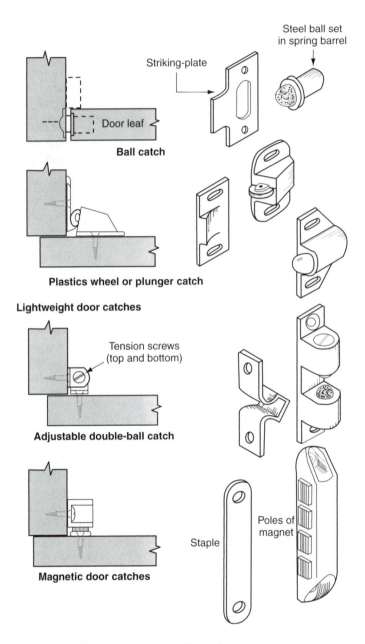

Fig. 10.22   *Cupboard catches*

10.5.2.1 Bottom-track sliding-door systems

Figure 10.26 shows examples of how some of the bottom-track systems work. Arrangements at the top are such that they keep the door from deviating from its straight path. This is achieved by pinning guide-laths on each side of the door, or using a channel to house the top edge of the door, or a retractable or non-retractable guide-pin.

Bottom arrangements include the use of nylon saddle slides (two per door) to carry the door over the top of a 'fibre' track, which is either

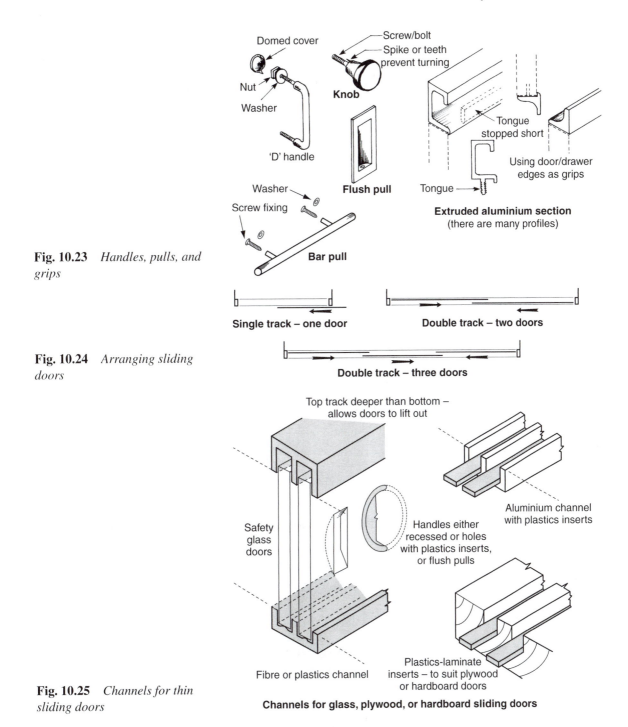

Fig. 10.23 *Handles, pulls, and grips*

Domed cover
Nut
Washer
'D' handle

Screw/bolt
Spike or teeth prevent turning
Knob

Flush pull

Washer
Screw fixing
Bar pull

Tongue stopped short
Tongue
Using door/drawer edges as grips

Extruded aluminium section
(there are many profiles)

Fig. 10.24 *Arranging sliding doors*

Single track – one door

Double track – two doors

Double track – three doors

Top track deeper than bottom – allows doors to lift out

Safety glass doors

Handles either recessed or holes with plastics inserts, or flush pulls

Aluminium channel with plastics inserts

Fibre or plastics channel

Plastics-laminate inserts – to suit plywood or hardboard doors

Fig. 10.25 *Channels for thin sliding doors*

Channels for glass, plywood, or hardboard sliding doors

housed in or surfaced fixed. Alternatively, wheel runners can be used over the same type of track. One problem with this arrangement is that the base track stands proud of its base board – restricting cleaning. This can in part be overcome by leaving the track short at each end.

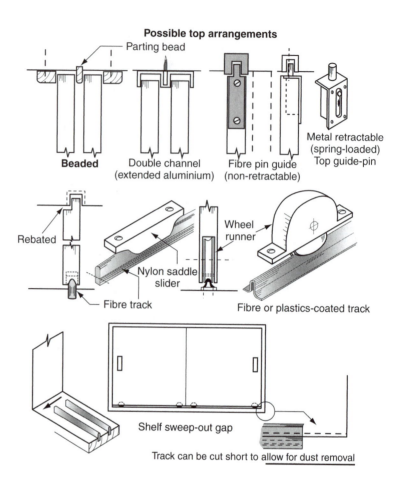

Fig. 10.26 *Bottom-track systems for lightweight cupboard doors*

10.5.2.2 *Top-hung sliding-door systems*

There are many excellent top-hung track systems on the market which permit doors to slide with the minimum of effort – one of the lightest of these is shown Volume 2, Fig. 9.9.

Figure 10.27 shows a 'Slik' wardrobe twin-track sliding system. Each door is hung from an aluminium track via two nylon wheels mounted on to a face-fixing bracket. The foot of each door is guided by small nylon semi-concealed upstands. A rubber buffer serves as a door stop.

Figure 10.28 shows a 'Slik' folding door gear system which enables doors to be slid and folded back at the same time, thereby permitting full access to the door opening. The doors are hung on their hanging edge by a top and bottom pivot. The top of the closing edge slides along an overhead track via pivot brackets set within it.

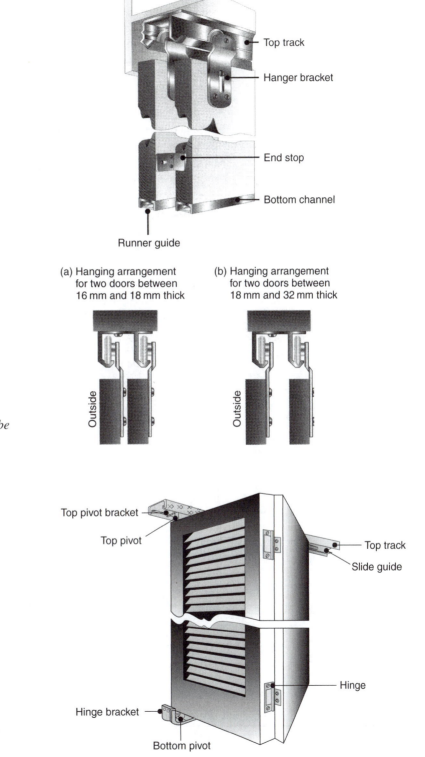

Top track

Hanger bracket

End stop

Bottom channel

Runner guide

(a) Hanging arrangement
for two doors between
16 mm and 18 mm thick

(b) Hanging arrangement
for two doors between
18 mm and 32 mm thick

Outside

Outside

Fig. 10.27 *'Slik' wardrobe twin-track sliding system*

Top pivot bracket

Top pivot

Top track

Slide guide

Hinge

Hinge bracket

Bottom pivot

Fig. 10.28 *'Slik' folding door gear*

10.6 Kitchen units

Kitchen units nowadays are available ready assembled (reducing installation time) in modular sizes to suit most kitchen appliances.

A method of installing the kitchen units is listed below and illustrated in Figs 10.29, 10.30 and 10.31.

(i) *Datum line (Fig. 10.29)* A horizontal datum line (see Datums, Volume 2, 1.6) is marked around the wall approximately 1 m above the floor where the kitchen units are to be fitted. As the floor may be out, it may be an idea to take a series of measurements to determine the highest point (or shortest distance to the datum line) of the floor. As appliances may be fitted into or under the units, measure the height of the units from this point. Remember that a refrigerator will need a ventilation gap at the top.

Adjustable legs are included with most modern units and are adjusted to suit varying floor levels.

The required height for the units is marked on the wall (measured at the floors highest point) and transferred around the wall, parallel to the existing datum line.

(ii) *Worktop back support rail (Fig. 10.30)* This rail (25×50) is secured to this new line with plugs and screws, and will provide a support to the worktop as well as a levelling point for the units.

(iii) *Wall units (Fig. 10.31)* At this point mark lines around the wall to position the units, from the datum line (Fig. 10.29), allowing

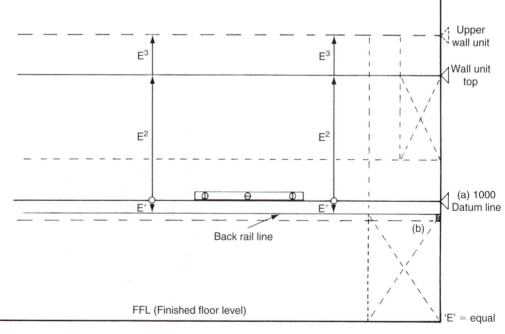

Fig. 10.29 *Fixing datum lines*

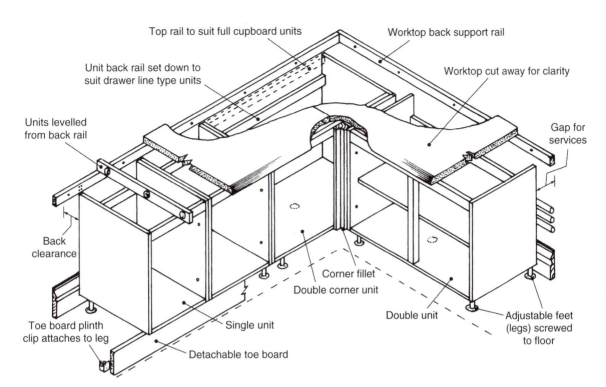

Fig. 10.30 *Fixing worktop base units*

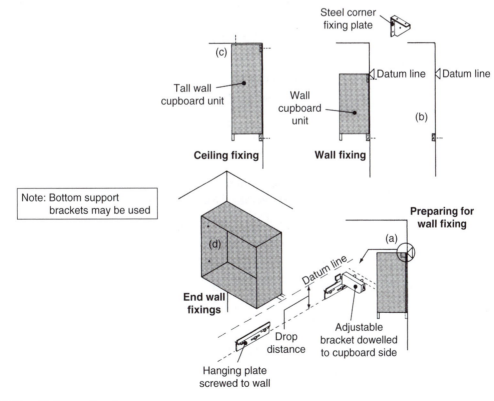

Fig. 10.31 *Fixing wall units*

for the thickness of the worktop and ensuring adequate headroom. If tall base units are to be included then these will determine the height of the wall units, if not, then it may be worth considering fitting the wall units first, starting in the corner, if corner units are included.

Wall units maybe levelled and secured onto timber battens with corner fixing plates (Fig. 10.31(b) and fixed to the ceiling (Fig. 10.31(c)).

Proprietary adjustable fixings are the modern alternative as shown in Fig. 10.31(a) end wall fixings shown at Fig. 10.31(d).

(iv) *Base units (Fig. 10.30)* Starting from the corner (if corner units are included) position the units in place and level off from the back support rail adjusting the leg as required. Allowance at the back of the units provides for installation of services etc.

All base and wall units are secured together with proprietary screws or bolts as shown in Fig. 10.12b.

10.6.1 Worktops

After the units are fitted and levelled, the worktop may then be placed in position to be joined (see Fig. 10.32).

This operation is carried out using a heavy duty electric router, template, guide bush (see Volume 1, p. 190, Fig. 6.44), suitable straight router cutter, and proprietary jig with locating pins which are secured in place using small 'G' cramps – Fig. 10.33.

Sufficient personal protective equipment (PPE) must be worn during the cutting process and an adequate extraction system fitted to the router.

Figure 10.32 shows the corner of the worktop that has been prepared prior to assembly, including also the slots for the panel connectors and biscuits for jointing and locating the tops together securely.

The necessary shaping of the joint is performed with a proprietary jig, for example the 'TREND' worktop jig as shown in Fig. 10.33.

This versatile jig, together with an electric router and guide bush, can be used to perform a number of required operations for different worktop layouts using the locating pins.

When using a proprietary jig the manufacturer's instructions must be followed to ensure a good quality joint on the worktop.

Figures 10.33 to 10.35 show the stages of operations involved when cutting and fitting the worktop.

After marking out the areas for cutting, the jig, which has been securely fitted to the edge of the worktop, is used with the router to perform the necessary shaping as shown in Fig. 10.33.

Worktop jig

Locating pins

Fig. 10.32   *The 'TREND' work-top jig and accessories*

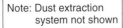

Note: Dust extraction
      system not shown

Worktop jig

Fig. 10.33   *Using the jig to form the joint*

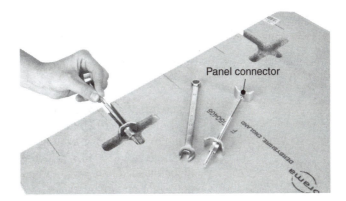

Panel connector

Fig. 10.34   *Fitting the panel connectors to the underside workshop*

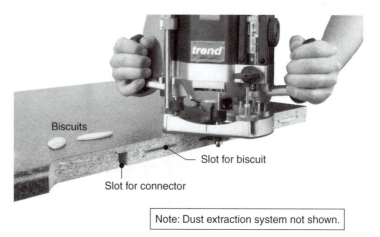

Biscuits

Slot for biscuit

Slot for connector

Fig. 10.35   *Forming the slots for biscuit jointing*

Note: Dust extraction system not shown.

The jig also includes facilities for forming the slots in each end of the worktops to receive the panel connectors. After the worktops are jointed together using a suitable adhesive, the panel connectors are then tightened using the appropriate spanner.

Figure 10.34 shows how the panel connectors fit into the panel slots.

The last operation is to cut the slots to receive the oval shaped biscuits that fit into the slots formed at both ends of the worktops. Once the adhesive has been applied, they will swell up and ensure a tight fit within each pre-formed slot.

A special self-guiding cutter is used with the router to form the slots for the biscuits (Fig. 10.35). Alternatively, a biscuit jointer could be used.

10.6.1.1 Cutting apertures in worktop

Forming the aperture/s in the worktop for sink units or cooking hob may be undertaken by the use of a jigsaw (see Volume 1, Chapter 6) or router (if a template is made, see Volume 1, 6.15.3) after first carefully marking out the correct positions. Masking tape may be used on the worktop for clarification of the marking out and to help prevent breaking-out of the cut edges.

Always double-check marking out otherwise mistakes cannot always be rectified after.

Figure 10.36 shows the cutting of an aperture by the use of a jigsaw after first drilling 12 mm pilot holes in each corner as starting points. By using a special down-cutting blade, break-out from the upper surface can be reduced. A firm grip and downward pressure on the jig-saw is recommended, otherwise there will be a tendency for the jigsaw to lift during the cutting operation.

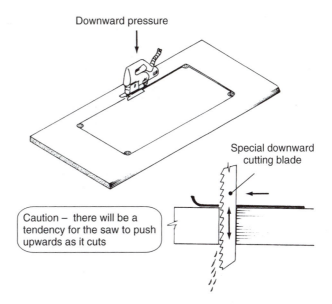

Downward pressure

Special downward
cutting blade

Caution – there will be a
tendency for the saw to push
upwards as it cuts

Fig. 10.36 *Worktop cut-outs*

All edges of the aperture must be sealed with an appropriate water-proof sealant.

10.7 Finishes

When all the units and worktop are fitted, the final operation is to fix the finishes to the wall units. This may be in the form of cornices or pelmets as shown in Fig. 10.37.

(i) *Cornices* – bind the units together at the same time providing a continuous decorative finish across the units. Usually available in 3 m lengths, either as a moulding of solid timber, or MDF covered with a decorative plastic laminate.

Using a purpose made mitre box, proprietary mitre box and fixed saw or electric powered mitre saw – mitres are cut to the required angle.

Mouldings must be held securely in the mitre box either by a temporary screw or turn button.

To prevent break-out to the exposed edges, all cuts must be made to the decorative face.

Particularly with solid timber, internal mitres are scribed (see Fig. 8.21) to minimise any shrinkage.

(ii) *Pelmets* – has similar functions to cornices by binding the units together and providing a decorative finish, but also may conceal electric lighting. Corners are mitred in a similar fashion to cornices and fixed to the underside by means of various methods.

To prevent light leakage along the joints an opaque sealant is used.

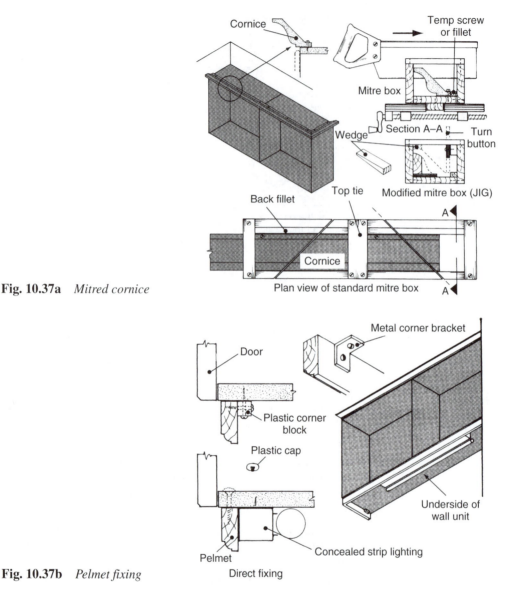

Fig. 10.37a *Mitred cornice*

Fig. 10.37b *Pelmet fixing*

References

Health and Safety at Work Act (HSAWA) 1974.
Control of Substances Hazardous to Health Act (COSHH) 1994.
BS 952–1:1995, Glass for glazing.
BS 4071:1996 (amended 1992): Specification of polyvinyl acetate (pva) emulsion for wood.
BS 6206: 1981: Specification for compact requirements.
BS 6262–4: 1994, Glazing for buildings.
Provision and use of work equipment regulations (PUWER).
Workplace (Health Safety & Welfare) Regulations 1992:

11 Shoring buildings

Shoring is a means of providing temporary support to a building being altered or repaired, and is used when a wall is in danger of moving and becoming unstable – possibly as a result of:

(a) natural subsidence (earth movement);
(b) mining subsidence (underground workings);
(c) undermining of foundations by water;
(d) lack of/breakdown of lateral restraint (bulging or overturning);
(e) structural alteration to the fabric of the building;
(f) vibration from heavy traffic, etc.;
(g) gale damage;
(h) an explosion;
(i) decomposition (erosion) of the building material;
(j) corrosion of the building material or part of it.

Flying shores are outside the scope of this book, but are easily identified because of the way they use adjacent buildings for support, for example across a street or over a courtyard.

11.1 Design

The design for shoring is undertaken by a qualified structural engineer who will determine the type of shoring and sizes of timbers and steelwork required. The sizes of components indicated in this chapter should be used as a guide only.

Consideration will be given to the layout of the site in relation to the amount of space available and the protection of the general public – to include temporary site hoarding, covered walkways (see Volume 2, section 2.2), and to identify and protect any existing underground services.

In a number of cases the shoring may be erected and left in place for a number of years before any work or alterations are carried out. This is often the case with 'raking' and 'flying shores', but the use of 'dead' shoring often involves the fitting or replacement of a lintel, or for an opening to be formed in an existing wall.

11.2 Types of shoring

Note: *In some cases it may be a combination of the above*

These can be in the form of:

(a) Raking shores – supporting buildings (Fig. 11.1a);
(b) Dead shores – supporting brickwork/stone above while undertaking work below (Fig. 11.1b);
(c) Flying shores – support between two buildings – nowadays usually in steelwork (Fig.11.1c).

11.2.1 Raking shores

A raking shore is used where a wall is at risk of moving outwards. Situations which could contribute to this effect have already been mentioned – several are illustrated in Fig. 11.2.

Although these shores are mainly used as a temporary means of support, there are instances – especially within inner cities – where they have remained erected for many years.

Ideally, rakers should be inclined at an angle of between 45° and 50° to the horizontal, but – because floor space is often limited by a boundary or restricted by an obstruction – this angle is often increased to a maximum of 60°.

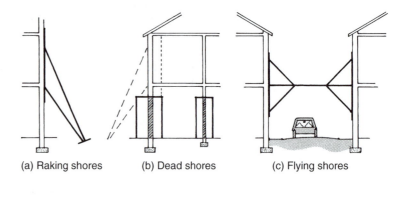

Fig. 11.1 *Types of shoring*   (a) Raking shores   (b) Dead shores   (c) Flying shores

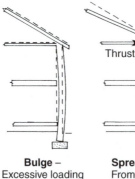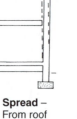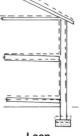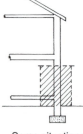

Fig. 11.2 *Situations requiring raking shores*

Bulge –
Excessive loading and/or settlement

Spread –
From roof (push over)

Thrust

Lean –
Due to settlement (subsidence, undermining, etc.)

Some situations associated with dead shores

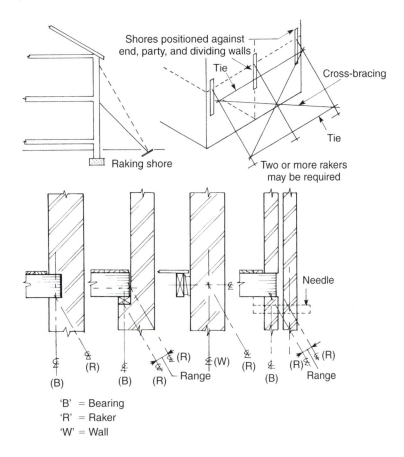

Fig. 11.3 *Positioning rakers*

'B' = Bearing
'R' = Raker
'W' = Wall

Figure 11.3 shows where and how rakers are positioned to counter any horizontal thrust being exerted from or about floors and/or walls. The required intersections of centre lines under different floor-to-wall situations are also shown.

11.2.2 Assembly and erection of a single raking shore

The following should be read in conjunction with Figs 11.4 and 11.5. Figure 11.4 illustrates the procedure, while Fig. 11.5 shows clearly the details of the shore and its components. Possible stages of erection are as follows (Fig. 11.4):

1. Locate and mark the position of the needle.
2. Mark the needle height and position on a rod.
3. Determine the length of the raker by:
    (a) scaling off a drawing (approximate);
    (b) using Pythagoras theorem to find the length of the hypotenuse (raker), see Volume 2, section 1.3.1;
    (c) trigonometry;
    (d) site measurement (where space permits): lay the rod on the ground, from or against a wall; form a triangle of the

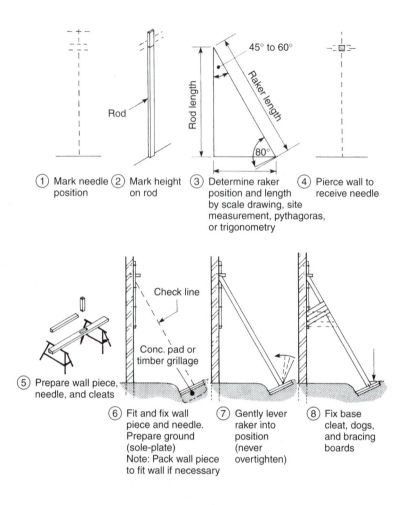

① Mark needle   ② Mark height   ③ Determine raker   ④ Pierce wall to
position   on rod   position and length   receive needle
                                by scale drawing, site
                                measurement, pythagoras,
                                or trigonometry

⑤ Prepare wall piece,   ⑥ Fit and fix wall   ⑦ Gently lever   ⑧ Fix base
needle, and cleats   piece and needle.   raker into   cleat, dogs,
                            Prepare ground   position   and bracing
                            (sole-plate)   (never   boards
                            Note: Pack wall piece overtighten)
                            to fit wall if necessary

Fig. 11.4 *Assembly and erection of a single raking shore*

Note: *Where more than one raking-shore system is used, they should be longitudinally tied at the top and bottom and cross-braced diagonally*

appropriate angle; then measure the length of the hypotenuse (raker) off the ground.

Order timber for the raker (a special order due to its large sectional size) or fabricate (laminate) it from smaller sections.

4. Cut a hole (100 mm × 75 mm) into the wall to a depth of not less than 150 mm to receive a hardwood needle (100 mm × 100 mm reduced to 100 mm × 75 mm at one end).

5. Prepare the wall piece, needle, and cleats.

6. Fit the wall piece to the wall – scribe and pack off as necessary. Fix it to the wall with wall hooks, anchor bolts or holdfasts (see Figs 11.5 and 12.7). Fit and fix the needle and cleats.

Prepare the ground for a concrete pad or timber grillage, then position and fix the sole-plate.

Check the raker length with a rod.

7. Prepare the raker and fix it by gently levering into position with a crowbar. (Never over-tighten!)

8. Fix the cleat to the sole-plate. Fix the raker to the sole-plate with a metal dog. Fit and fix bracing boards.

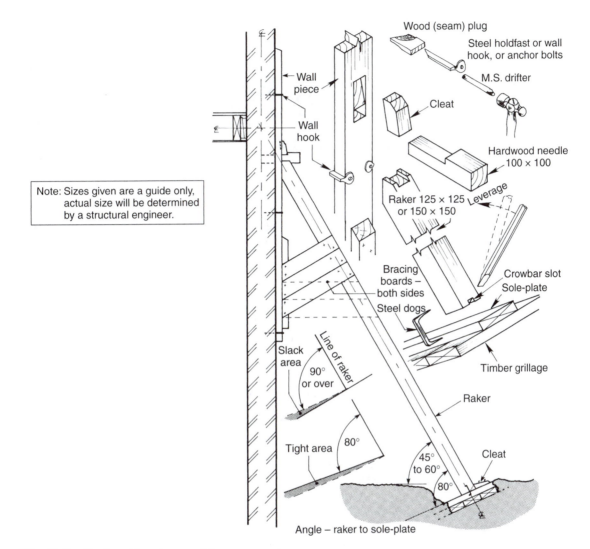

Note: Sizes given are a guide only, actual size will be determined by a structural engineer.

Fig. 11.5 *Single raking shore and its components*

11.2.3 Dead shores

Dead shores are used to support the walls of the building and any load that a wall may endure, for example the self-weight of the wall and any other permanent fixed parts of the fabric (floors, roofs, etc.) which rely upon it for their support. These are known as 'dead' loads. When the need arises, certain live loads (imposed loads) such as occupants and furniture, etc. must also be carried by the dead shores. Dead shores come into operation when part of a wall is about to be removed for reasons of repair or alteration. If the work involves shoring-up (propping) a wall so that work can be carried out on its foundations, this type of shoring is generally known as 'underpinning'.

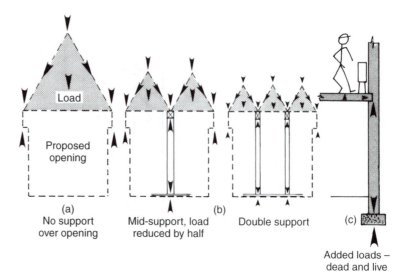

Fig. 11.6 *Taking into account loads over proposed openings*

Figure 11.6 shows how loads above an opening must be taken into account when deciding upon:

(a) the opening length;
(b) the number of shores and the method of propping;
(c) added dead and live (imposed) loads.

Foundations to the side of the proposed opening must be strong enough to carry any extra loading after the opening has been formed.

If ever there is a danger that disturbance to the wall might render it unstable, raking shores should be included in the overall shoring design.

11.2.4 Erection procedures

What follows refers in part to the illustrations shown in Fig. 11.7. However, Figs 11.8 (shoring an interior load-bearing wall) and 11.9 (shoring an exterior wall) should also be referred to with regard to detailing and because location of the shores can differ.

(i) Assess the situation with regard to dead and imposed loads (Fig. 11.6).
(ii) Strut any openings (e.g. window openings) within the wall above and to the sides of the section which is to be removed (Fig. 11.9a).
(iii) Using timber or adjustable steel props (Volume 2, Figs 3.20 and 4.17), prop the floors and roof above and about the proposed opening (Fig. 11.7a). Propping should be continuous from a load-bearing base up to the roof structure.
(iv) Pierce the walls to receive needles at about 1.2 m centres depending on the condition and type of wall material (Fig. 11.7b). Holes should not be positioned under window openings.

(v) If access to the opening is to be restricted once the shores are in position, place the new lintel (beam) – a steel universal beam (UB) or reinforced concrete – against the foot of the wall at this stage.

(vi) Position needles (steel or timber, Fig. 11.10) through the wall, after making sure that when the shore is in position they are long enough to allow a working space to one or both sides of the existing wall (Fig. 11.7c). The material and sectional size of the needles will determine their permitted overhang.

(vii) Position sole-plates to both sides of the wall upon a solid base.

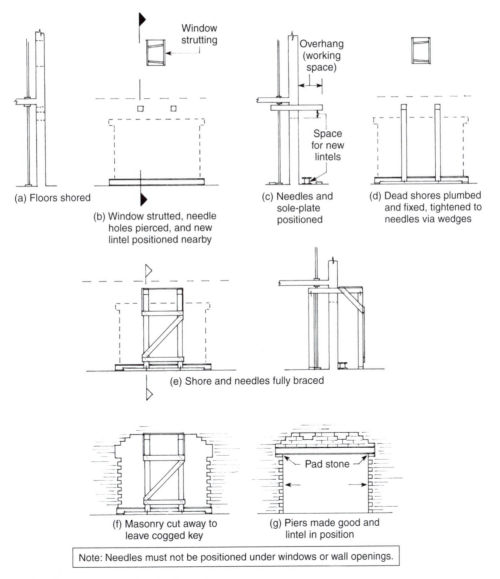

(a) Floors shored

(b) Window strutted, needle holes pierced, and new lintel positioned nearby

(c) Needles and sole-plate positioned

(d) Dead shores plumbed and fixed, tightened to needles via wedges

(e) Shore and needles fully braced

(f) Masonry cut away to leave cogged key

(g) Piers made good and lintel in position

Note: Needles must not be positioned under windows or wall openings.

Fig. 11.7   *Erection procedures for dead shoring*

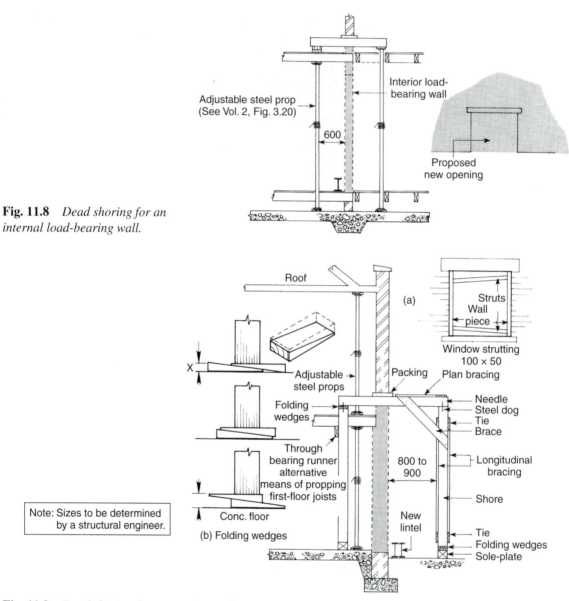

Fig. 11.8 *Dead shoring for an internal load-bearing wall.*

Adjustable steel prop
(See Vol. 2, Fig. 3.20)

Interior load-
bearing wall

600

Proposed
new opening

Roof

(a)

Struts
Wall
piece

Window strutting
100 × 50

Adjustable
steel props

Packing

Plan bracing

Folding
wedges

Needle
Steel dog
Tie
Brace

Through
bearing runner
alternative
means of propping
first-floor joists

800 to
900

Longitudinal
bracing

Shore

Conc. floor

New
lintel

Tie
Folding wedges
Sole-plate

(b) Folding wedges

Note: Sizes to be determined
by a structural engineer.

X

Fig. 11.9 *Dead shoring for an exterior wall*

(viii) Position dead shores (Fig. 11.7d). Sectional size will depend on:
  • the type and height of the building (the number of storeys);
  • the length of the shore;
  • the distance between shores.

They should provide a stable base for the needles (see Fig. 11.8).

After plumbing the shores, use hardwood folding wedges (Fig. 11.9b) to tighten the needles up against the underside of

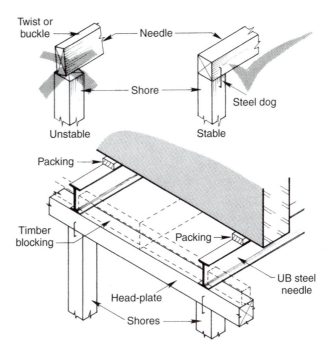

Fig. 11.10 *Providing a stable base for needles*

the wall. Packing may be needed to ensure that the needles are level. Wedges may be positioned at the head or foot of the shore.

Secure the joints with cleats (nail heads left to draw) and steel dogs.

(ix) Brace the needles to the shores and longitudinally brace and tie the shores together (Fig. 11.7e). Plan bracing (across the top of the needles) may also be used.

(x) Check all joints inside and out for tightness.

(xi) Masonry may now be carefully removed as necessary (Fig. 11.7f).

(xii) Rebuild piers to form a bearing for the new lintel (Fig. 11.7g). The pad stone is usually cast in concrete. Construct formwork if an *in-situ* concrete lintel is to be cast (see Volume 2, Fig. 3.19). After the lintel is installed and/or set, the masonry above is made good.

(xiii) Leave the masonry until it is set and strong enough to sustain the loads to be placed upon it when the shoring is removed.

(xiv) After the prescribed period of time, the shoring is 'eased' (by gently slackening the folding wedges a small amount – any settlement of the wall structure will be detected if the wedges start to retighten) and eventually struck (dismantled). This stage should be undertaken only on the instruction and with the supervision of the person responsible for the design and construction of the shoring – usually a structural engineer.

(xv) Dismantle the shoring in reverse order to its assembly.

(xvi) If raking shores have been used to accompany the dead shores they will be dismantled last.

11.2.5 Flying shores

This type of shoring (Fig. 11.1(c)) is used in situations where, for example, an opening has been formed between buildings, which may include a row of terraced houses. The remaining end walls therefore must be supported to prevent distortion, which will lead to eventual collapse. Generally, the shoring is situated high enough to allow maximum support and, if required, pedestrians or traffic.

Shores of this nature are specifically designed by structural engineers.

11.3 Safety

Because shoring is required to support and/or restrain a load-bearing structure, its design is very important. In the interests of safety, therefore, all shoring systems should be designed by a structural engineer (including calculations to show the loading upon all shoring load-bearing members and the sectional sizes and spacing to support those loads) and should be erected by persons competent to do so, under the direction of the designer (usually a structural engineer).

Safe working platforms must be provided for operatives erecting, servicing, and dismantling shoring – see Volume 2, Chapter 10 'Scaffolding'. Many shoring operations will require the erection of a hoarding as a means of offering site and public protection, see Volume 2, Chapter 2 'Fences and hoarding'.

References

Construction (Health, Safety and Welfare Regulations) 1996.
HSE Information Sheet. General access, scaffolds and ladders – Information sheet 49.
Heath and Safety Regulations 1977.
The Highways Act 1980.
BS 5268–5: 1989, Structural use of timber, code of practice for the preservative treatment of structural timbers.
BS 5973: 1993, Code of practice for access and working scaffolds and special scaffold structures in steel.
BS 1139–3: 1994, Specification for prefabricated access and working towers.
BS 1139–4: 1982, Specification for prefabricated steel splitheads and trestles.
BS 1129: 1990, Timber ladders, steps, trestles, and lightweight stagings.
BS 2037: 1994, Aluminium ladders, steps, and trestles for the building and civil engineering industries.
BS 2482: 1981, Specification for timber scaffold boards.
BS 5973: 1993, Section 9 – tying to buildings.

12 Repairs and maintenance

During their working life, most carpenters and joiners will spend a great deal of their time repairing and/or maintaining various items associated with buildings and their contents.

Firms that specialize in this type of work tend to employ craftsmen who are flexible, in the sense that they are conversant with and capable of carrying out basic jobs outside their own specialist area. For example, a carpenter and joiner may be expected to glaze or re-glaze a window, or use a trowel and float to do some patchwork on a plastered wall or concrete floor. The ability to lay or renew broken tiles or slates may also be a requirement.

A sound knowledge of building construction is essential for all skilled operatives engaged on this type of work.

Certain items of hardware are expected to be replaced periodically, due to wear through use; failure to provide and carry out a general maintenance programme (for example, to ensure that the building remains watertight and that protective coatings to woodwork are regularly reapplied, etc.) will lead to deterioration and eventual failure of many other building materials.

In these days of short-term apprenticeships, multi-skill training to a high standard is not practicable. This, among other factors, has led to the setting up of firms which specialize in specific areas of work, for example timber decay due to fungal or insect attack, damp-proofing, insulation, and replacement windows, to mention just a few.

12.1 General repairs survey

As with any job, the way in which one organizes repair work is very important; therefore, when called upon to survey a building with regard to carrying out repairs, a true record should be made of its existing condition. There are four ways by which a record can be made:

  (i) a detailed list (schedule),
 (ii) annotated sketches,
(iii) a combination of (i) and (ii),
(iv) camera.

Figure 12.1 and Table 12.1 have been drawn up as an example of how a pre-war 'back-to-back' end terrace house could be surveyed by

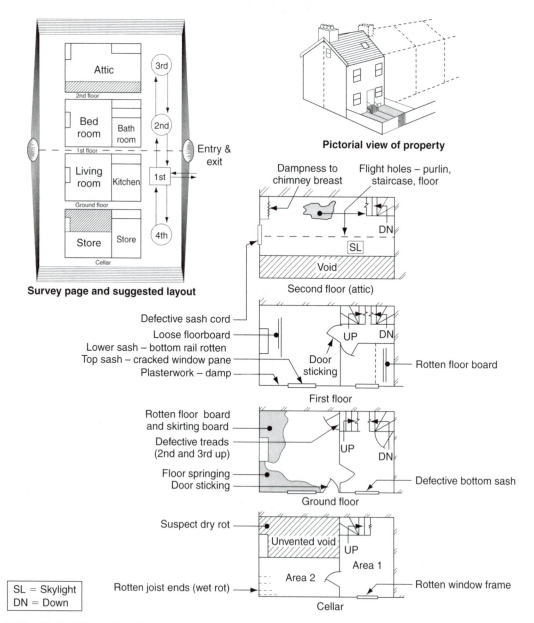

Fig. 12.1 *Typical floor plan layout to show some possible defects to a pre-war back-to-back terrace house*

Table 12.1 Example of how a repairs survey on a back-to-back end terrace house could be tabulated (to be used with Fig. 12.1)

Location	Defect	Possible cause	Remarks
1. Living room	Rotten floorboard and skirting board	Dry rot (unvented under-floor space)	See Volume 1, Chapter 2
	Floor area under window springing	Cellar rot (wet rot) – confirmation in cellar	See Volume 1, Chapter 2
	Exterior door sticking	Over-painted; moisture movement; door sagging; defective hinge; etc.	See Table 12.2
2. Stairs to first floor (second and third treads)	Nosing split	Heavy wear; timber defect; riser groove cut too deep.	See section 12.4.9
	Squeaking steps	Shrinkage due to moisture movement; treads and risers unhoused.	See section 12.4.9
3. Bathroom	Rotten floor board under bath	Wet rot, due to water spillage or leaking waste/water pipe.	See Volume 1, Chapter 2
4. Bedroom	Defective door latch	Worn and over-painted	Renew
	Door sticking	Defective hinge	Renew
	Loose floorboard	Access trap to services	See section 12.4.8
	Cracked window pane (top sash)	Unknown	Reglaze – see section 12.4.5
	Rotten bottom rail (lower sash)	Window condensation	Renew
	Damp plasterwork	Surface condensation – penetrating damp	See Volume 1, Chapter 2
5. Stairs to second floor (attic)	Flight holes in treads, risers and strings	Woodworm (common furniture beetle)	See Volume 1, Chapter 2
6. Second floor (attic)	Flight holes in the purlin and floorboards	Woodworm (common furniture beetle)	Volume 1, Chapter 2
	Gable-end window sash will not open	Defective sash cords	Renew all cords – see section 12.4.7
	Dampness to chimney breast	Defective cement pointing, lead step, flashing, or lead soakers; interstitial condensation.	See Volume 1, Fig. 2.5
7. Note: On returning downstairs, recheck rooms and staircases.			
8. Kitchen	Window (bottom sash) – joint of stile and bottom rail	Defective haunch or wet rot	Dismantle sash, insert false tenon, possibly splice stile
9. Cellar (area 1)	Rotten window frame	Cellar rot (wet rot)	See Volume 1, Chapter 2
(area 2)	Rotten joists ends (living room floor)	Cellar rot (wet rot)	See Volume 1, Chapter 2
Cellar void	Dampness (see living room – dry rot)	No ventilation	Provide ventilation (air bricks) to external wall – pierce interior walls

using method (iii). – Further examples may be found in Work Activities in Section 13 by Brian Porter and Reg Rose.

The listing of external repairs is often dealt with separately – preferably after the internal inspection. This is because defects found within the dwelling could be the result of an exterior defect which, without close scrutiny, might otherwise have been missed. For example, a damp patch on the inner face of an external wall could be the result of surface condensation, but on the other hand it could have been caused by an excessive amount of rainwater running down the outside face wall, due to a blocked fall pipe or a defective gutter joint.

Provisional observations from ground level can be enhanced by the use of binoculars. Or a more detail report may need the use of some form of scaffolding (see Vol 2, chapter 10 Scaffolding).

A survey of this nature should cover the following:

> Note: *A similar survey for a semi-detached can be found in* Work Activities, *section 13*

1. *Face walls*
   - Stability (any signs of structural movement – settlement cracks) – note the close proximity of any trees.
   - Height of the damp proof course (DPC) above the ground level – should not be less than 150 mm.
   - Air grate (air bricks) clear and away from any overgrown vegetation.
   - Cement pointing – no open joints.
   - Cement rendering – no visible cracks or loose parts (no hollow sound when gently tapped).
   - Staining (no undue discolouration) – particularly just above the DPC line or under the roof eaves, and behind the rain water fall pipes.

2. *Wall openings*
   - Doors – general condition of the woodwork, and particular attention to weather seals around the frame, threshold and weatherboard.
   - Windows – general conditions of the woodwork, glass and putty/sealant pointing – particularly weather seals around window frames.

3. *Fall pipes*
   - Soil pipes – joints fully caulked (sealed).
   - Vent pipes – joints fully caulked (sealed), and extend above any window or house vent opening at least 900 mm if less than 3 m away (as per Building Regulations).
   - Particular attention should be paid to the backs of cast iron pipes, as this is usually the first place to corrode (rust) due to lack of paint protection.

4. *Gully grates*
   - Check for signs of water being discharged over the edges into the soil, onto the path/patio or up against the wall of the house.

- Check the water level in the 'U' bend by pouring a bucket of water down into the gully.

5. *Eaves guttering (spouting)*
   - Falls – no signs of static water in the guttering. Test the fall of the guttering on a dry day by pouring water into the guttering at its highest point.
   - Joints – check for any water stain on the underside of the guttering, and signs of any moss growth.
   - Vegetation growth within the gutter – due to static water or poor falls. Check any wooden gutters for similar signs, which may also be due to similar conditions or possible decay.

6. *Eaves (pitched and flat roof)*
   - Fascia boards (open, closed or flush eaves) – for signs of decay and/or distortion.
   - Soffit boards (closed eaves) – for, the presence of any ventilation grills or gaps.

7. *Flat roofs*
   - Surfaces – visible falls, signs of ponding (retained grit residue on roof surface without solar chippings), check also for any signs of vegetation growth.
   - Solar chippings – in place on built-up felt roofs.
   - Flashings and aprons (metals (lead) or felt) – intact and securely fixed within their abutment wall.

8. *Pitched roofs*
   - Surfaces – check for missing, displaced or defective roof tiles and slates. Also missing displaced or defective hip and ridge tiles.
   - Verges – cement pointing to verge tiles/slates. Check the roof finial, bargeboards and soffits for signs of decay and/or distortion, etc.
   - Roof lights (skylights) – glazing and water seals.
   - Dormer windows – as above (roof surfaces, eaves, glazing and water seals).

9. *Chimney stacks*
   - Brickwork, blockwork, etc. – stability (no signs of structural movement), cement pointing or rendering is intact – check for any sooty deposits between joints, in particular at perpends (vertical joints).
   - Flaunching (cement collar around the chimney pots) – intact, check also for sooty deposits at joints with brickwork and chimney pots.
   - Chimney flashings – check for missing or defective step flashings and soakers ('L' shaped pieces of metal – usually lead), which lap one another from the chimney back gutter to the front apron to form a weather seal between the underside of the stepped flashings and roof tiles/slates. The front apron can

often be viewed from ground level. However, the back gutter would require access to the roof.

10. *House steps*
    - Check for level and surface wear. Any cracks, mould growth etc.
    - Check handrails for stability, fixing and fully intact.
    - Does the stair comply with current Building Regulations – Approved Document 'K' in relation to pitch, rise or going, headroom, handrails or balustrade?

11. *Garden path and driveway*
    - Surface condition – cracks, mould growth etc.
    - Drain inspection chamber – metal covers intact and secure.

12. *Boundaries*
    - Walls – make observations with regards to settlement and distortion (The local authority must be informed of any walls adjacent to a public thoroughfare or highway that appears to be in a dangerous condition).
    - Wooden fences and gate posts – check for any signs of decay particularly at and around ground level and where water can be retained, such as where rails cross over posts etc. (see Volume 2, 2.5).
    - Wooden gates – check for sag (see 4.2.2), distortion (twist), and decay, particularly at joints, junctions and where members overlap (as above), also behind metal fixings.

13. *Out-buildings (attached and/or detached)*
    - Garages – observations as above.
    - Porches and conservatories – as above.

14. *Dangerous materials and liquids*
    - Make a note of any suspicious materials or liquids (foul water etc.) that you may encounter. For example asbestos which was once a common building material used in roofing and pipe insulation etc. – seek advice where necessary.
    - Always wear the appropriate safety wear. Carry out a risk assessment before proceeding with the site survey.

Once the site survey has been completed, you may be required to pass on this information to a colleague via a written report. You could also include a simple annotated line diagram of each elevation indicating any defects. Include also a written schedule similar to Table 12.1 and any photographs of particular areas showing defects areas from the survey.

12.2 Customer relations

Most items of repair work are done while the building's occupants are in residence. The occupants should be inconvenienced as little as possible, and care must be taken to minimize damage to decorations and not to damage furniture in any way. Exterior work may involve working from or over a cultivated garden. Landscaping and planting-up of even a small garden can run to many hundreds, if not thousands, of pounds – therefore care must be exercised both inside and outside the building as necessary.

Occupants should receive prior notice of the nature and timing of the work proposed (time span will depend on the type and amount of work to be done) well before work begins to allow them time to prepare the site as necessary. Most firms are expected to provide dust-sheets and assistance in moving heavy furniture, lifting carpets, etc.

Operatives employed to carry out small interior repair or maintenance work within a fully furnished home should be prepared to wear the appropriate footwear.

Gardens should be treated with similar respect – a lawn could take years to recover from damage caused by the foot of a ladder, etc.

Exterior finishes to house walls can easily be damaged by rearing or running ladders, etc. up against them. If the wall surface is liable to become marked or damaged, consider wrapping the top of the ladder.

Never allow anything to be dropped to the ground, be it from a scaffold roof or through a window. The risk of injuring a passer-by or by-stander may be obvious, but the possible damage to the object and its landing place are not always apparent.

Finally, one cannot emphasise enough that from the outset of the job that a little care and consideration can go a long way towards good customer relations, which will not only provide you and your company, but will also add credits to your competence as a skilled capable operative.

12.3 Hand/power tools and equipment

Once an item of woodwork has been installed, fixed, and possibly serviced by other trades, its removal or repair and eventual reinstatement may require the use of tools and equipment outside the normal list generally prescribed for a bench-hand or a building-site worker. In fact many of the tools found in a maintenance joiner's toolbox or bass (basin-shaped stout hessian holdall) would normally be found in tool kits used by an engineer or bricklayer.

Listed below are examples of the types of hand and power tools required (see also Volume 1, Chapters 5 and 6), including equipment often needed. Items marked * may be carried in a Joiners bass.

Quality and sharp tools should always be protected. The main reason for using a bass is that it is easy and light to carry, and tools are instantly accessible. The main disadvantage is that unprotected tools can easily become damaged by the constant knocking of one against another. Items such as sharp chisels may protected by plastic caps or masking tape.

(a) Measuring, setting-out, and marking tools
    (i) 1 m folding rule;
   (ii) 3 m steel tape measure;
  (iii) 1 m spirit level;
  (iv) 300 mm spirit level (or boat level)*;
   (v) Builder's line (chalk line)*;
  (vi) Plumb bob*;
 (vii) Combination square*;
(viii) Marking gauge*;
  (ix) Sliding bevel*.

(b) Saws
   (i) Panel saw*;
  (ii) Tenon saw (short-bladed)*;
 (iii) Bushman-type tubular-steel frame saw;
 (iv) Flooring saw;
  (v) Pad saw*;
 (vi) Hacksaws – large and junior*;
 (vii) Coping saw*;

Saw teeth must be protected with a sheath if saws are to be carried in a bass. (Vol. 1, Chapter 5)

(viii) Jig saw;
  (ix) Reciprocating saw;
   (x) Circular saw.

See Volume 1, Chapters 6 and 7 for safe use of power tools

(c) Planes
  (i) Jack plane (wooden types are often preferred to metal, because they are lighter and less likely to break);
 (ii) Smoothing plane (metal)*;
(iii) Rebate plane*;
(iv) Block plane*;
 (v) 'Surform' types (planes and files)*;
(vi) Planer (see Volume 1, Chapters 6 and 7 for safe use of power tools).

(d) Boring tools
  (i) Carpenter's brace and an assortment of bits, including a countersink, and a turnscrew*. Drill bits with tapered square shanks (morse drills) and an expanding-bit set should be included. Loose bits should be kept in a stout bit roll; (Vol. 1, Chapter 5)
 (ii) Hand drill (wheel-brace), with a boxed set of twist drills*;
(iii) Set of bradawls*;
(iv) Drill/screwdriver (see Volume 1, Chapters 6 and 7 for safe use of power tools).

(e) Chisels*
An assortment of:

 (i) Bevel-edge chisels;
 (ii) Firmer chisels;
 (iii) Registered chisels;
 (iv) Gouges;
 (v) Engineer's cold chisel – for cutting steel and masonry*;
 (vi) Mason's chisels – for plugging (seaming chisel and star-drill chisels) and cutting (bolsters)*;
 (vii) Flooring chisel (tonguing tool)*.

All loose chisels must be kept within a stout chisel roll. (Vol. 1, Chapter 5)

(f) Axe or blocker*
The blade must be protected with a thick, strong, leather or similar sheath.

(g) Hammers
 (i) Engineer's hammer*;
 (ii) Club or lump hammer;
 (iii) Warrington hammer (number 2)*;
 (iv) Claw hammer*;
 (v) Nail and staple gun (see Volume 1, Chapters 6 and 7 for safe use of power tools).

(h) Sanders
 (i) Belt;
 (ii) Orbital sheet heavy duty
 (iii) Random orbital 'Palm' grip   } (see Volume 1, Chapters 6 and 7 for safe use of power tools);
 (iv) Detail sander.

(i) Screwdrivers
 (i) Rigid-blade types – large and small (electrician's) slotted and Pozidriv*;
 (ii) Ratchet*;
 (iii) Spiral ratchet, with an assortment of bits and drill points;
 (iv) Turnscrew (for use with a carpenter's brace)*;
 (v) Power screwdriver (see Volume 1, Chapters 6 and 7 for safe use of power tools).

(j) Pincers and pliers*

(k) Adjustable spanner*

(l) Punches
Nail punches, centre punch, and drifts*.

(m) Metal files
- (i) Mill saw file (with a wood/plastics handle) or a 'Farmer's friend' file with a metal handle*;
- (ii) Saw file;
- (iii) Warding file*;
- (iv) Needle files (set).

(n) Knifes
- (i) Pocket knife;
- (ii) Trimming knife*;
- (iii) Cobbler's hacking knife – for removing hard dried putty pointing (fronting) and bedding from glazed window rebates*;
- (iv) Putty pointing knife*.

(o) Trowels
- (i) Bricklayer's pointing trowel*;
- (ii) Mastic pointing trowel*;
- (iii) Plasterer's float.

(p) Ancillary equipment
- (i) Dust-brush (painter's)*;
- (ii) Oilstone (combination stone in box)*; (an assortment of sharpening stones are available)*;
- (iii) Oil can
- (iv) Cork block and abrasive papers*;
- (v) Wrecking bar (tommy bar);
- (vi) 'G' cramps (minimum two);
- (vii) Saw stool or 'Workmate'; (or similar)
- (viii) Nail and screw box;
- (ix) Nail and screw bag (circle of stout fabric with draw-string) for use with a bass – filled with assorted types and sizes*; (see Vol. 1, Chapter 5)
- (x) Window-cord box – containing a 'mouse', pulley pegs (pointed dowel), sash nails (tacks), and 38 mm oval nails;
- (xi) Panel-pin tin (box), assorted sizes*;
- (xii) Joiner's bass (note: a plumber's bass is smaller); (see Vol. 1, Chapter 5)
- (xiii) Toolbox;
- (xiv) Nail and screw pouch;
- (xv) Tool-belt and holsters;
- (xvi) Plasterers hand-board (hawk);
- (xvii) Mastic gun with assortment of cartridges;
- (xviii) Dustsheets (fabric or plastic);
- (xix) Box of plastic plugs (assorted sizes).

(q) Personal Protective Equipment (P.P.E.)
- (i) Respiratory protection
  There a number of different types of protective masks available, each suited for various working situations. For example working with:

- Clay products, hay, pollen, iron products
- MDF dust;
- Mist and fumes;
- Organic vapours, fine toxic dust and fibres and aqueous mists;
- Solid and liquid aerosol variants including brick and wood dust, concrete plaster and sandstone;
- Water based mists.

(ii) Ear defenders

These need to be comfortable as they may be worn for long periods of time. Various types include:

- Banded ear plugs;
- Ear plugs;
- Electronic ear-defenders;
- Foam ear plugs;
- Large foam-filled cups and adjustable headband.

(iii) Goggles/Specs/Full mask

Need to be impact resistant, anti-mist and allow the operative maximum visibility. Various types are available for protection against:

- Bright light that may damage eyes, i.e. welding;
- Grinding particles, sanding dust and pressure washing;
- Liquid droplets and molten metals.

(iv) Head protection (Safety helmet, Hard hat)

Provides protection to the head against falling or stationary objects. Made from high density polythene. Some types include ear and visor protection in one. Check the life span of the hard hat or bump cap and replace when necessary. Types include:

- Bump cap;
- Hard hat.

(v) High visibility clothing

A highly reflective material, which allows the operative to be seen whilst working in dangerous situations. Must be water proof and hardwearing. Available as:

- A jacket;
- Loose bands;
- Trousers;
- Waistcoat.

(vi) Knee protection

Protection of the knees is important, particularly when working on a hard, wet or uneven surface. Various types include:

- Anti-slip, no-mark knee pads;
- Gel swivel knee pads – in situations of continuous turning;
- Knee pads – general use;
- Kneeling mat;
- Safety trousers with protective knee pads.

(vii) Protective gloves (Safety gloves)
Various types are available which can be made of cowhide, latex and other synthetic materials, or a knitted material with various coatings. Depending on their use, they need to be hard wearing, heat, tear and puncture resistant. Some types may have extra padding on the palms and fingers and may include finger grips. Other types are disposable after one use. Protective gloves are used for protection against:
- Abrasions;
- Chemicals - acids and caustic corrosive chemicals;
- Cuts;
- Heat;
- Splinters.

(viii) Protective footwear (Safety boots)
There are many situations where the foot may be damaged by hard surfaces, heavy objects, protruding nails or chemical spillage. Footwear may include metal toe guards and/or sole protectors against penetrating items such as nails or other sharp objects. They may also be coated to resist corrosive liquids. Types include:
- Boots;
- Disposable over shoes;
- Shoes;
- Sports trainer boots;
- Trainers;
- Wellington boots.

Note: Under the Health and Safety Act 1974, all P.P.E. must be provided by the employer and worn where necessary.

(r)   Testing equipment
Various testing equipment may be used before any work commences i.e. location of services, and studwork checking a material for moisture content etc. These include:
- Moisture meter (see Volume 1, 1.7.1);
- Stud locator (see Fig. 8.43);
- Pipe/metal locator (see Fig. 8.43);
- P.A.T. (Portable Appliance Test) unit.

12.4 Remedial treatment

Any items of woodwork – structural or non-structural – could at some time be liable to damage or deterioration. Maintenance joiners should therefore be prepared to tackle almost any type of job – the following are just a few of the items that could be encountered.

Table 12.2 Remedial treatment for defective doors and door frames

Defect	Possible cause	Remedial treatment
Sticking (closing edge)	Build-up of paint. Moisture movement due to moisture intake (check with moisture meter)	*Ease closing stile (Allow for clearance) Seal/repair.†
Sagging (Dropped)	Design or construction fault – see Chapter 4, Section 4.2.2.	Refit (cut to shape) or reassemble door – see Fig. 12.2.
Binding (springing)	Build-up of paint over hinges or within hanging-stile rebate. Oversized screw heads to hinge. Bowed hanging stile.	Remove build-up of paint. Rehang with longer, smaller-headed screws. Bevel hanging edge of stile (see Fig. 12.3).
Twist	Irregular wood grain – subjected to variable atmospheric conditions; post-seasoning after door is fixed; construction fault; door frames twisted (see Volume 1, Table 1.7).	Depending on the degree of twist, see Fig. 12.4 or renew door.
Rotten or damaged stile and/or rails	Wet rot; accidental or criminal damage.	Cut away defect (Fig.12.5(a)) – use keyed splices and false tenons (Fig. 12.5(b)) as necessary.

* Remove old paintwork with a scraper (shaved-hook) or 'Surform' etc. (Always wear safety glasses or goggles during this hacking or scraping process). Plane down as necessary.
† Use micro-porous paint only on the edges of outward-opening doors and windows.

Probably the most common of all jobs, and often the simplest, is the easing of 'sticking' doors and windows. Door faults usually come to light after redecoration (build-up of paint) or in the autumn, when unsealed or unprotected timber is exposed to a prolonged increase in moisture content, causing it to swell. Sticking window casements are most likely to be noticed in the spring, when most of the casements are opened for the first time after being closed during the months of winter.

12.4.1 Door repairs

See Table 12.2.

Door frames, and indeed window frames if not properly protected or maintained will lead to the frame being repaired or at worst removed and replaced.

12.4.2 Door frame repairs (see Table 12.2)

Figure 12.6a shows the sequence of removing sections of a frame if it needs replacing. However, Fig. 12.6b shows how a rotten door

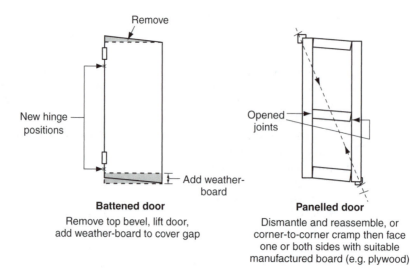

Battened door
Remove top bevel, lift door,
add weather-board to cover gap

Panelled door
Dismantle and reassemble, or
corner-to-corner cramp then face
one or both sides with suitable
manufactured board (e.g. plywood)

Fig. 12.2   *Sagging doors*

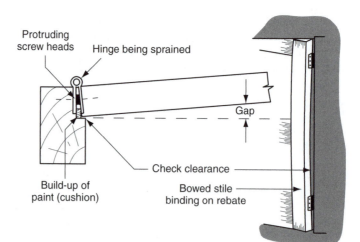

Fig. 12.3   *Hinge-bound doors*

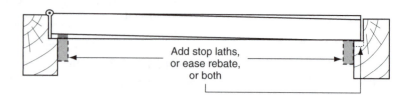

Fig. 12.4   *Possible remedy for
slightly twisted doors*

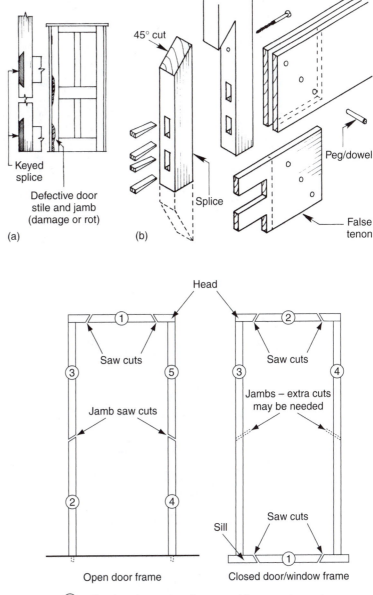

Fig. 12.5 *Repairing a damaged door*

(a)

(b)

Keyed splice

Defective door stile and jamb (damage or rot)

45° cut

Splice

Peg/dowel

False tenon

Fig. 12.6a *Removing door/window frames from their openings*

Head

①

②

Saw cuts

Saw cuts

③

⑤

③

④

Jambs – extra cuts may be needed

Jamb saw cuts

②

④

Sill

Saw cuts

①

Open door frame

Closed door/window frame

ⓧ = Numbered sequence for removal from opening

frame (usually wet rot or damaged jamb foot) can be repaired by splicing.

- Make 45° cuts sloping down towards the exterior – in this way, water is shed away from the joint (Fig. 12.6b(i)).
- Remove the defective portion. Expose and clean out the anchor peg hole (Fig. 12.6b(ii)). Use a sharp chisel to chop away waste when forming the scarf.

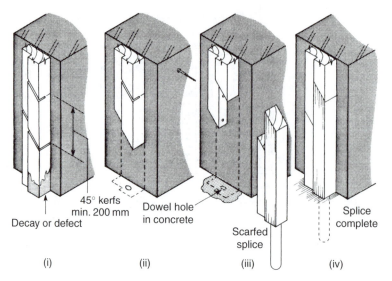

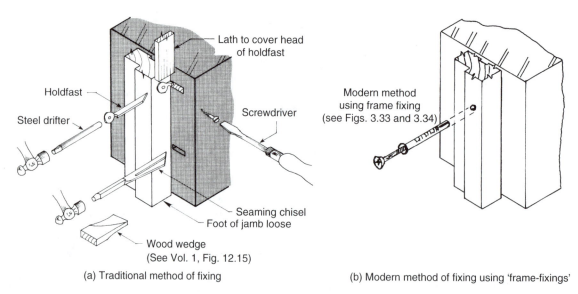

Fig. 12.6b *Splicing a damaged door jamb*

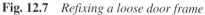

Fig. 12.7 *Refixing a loose door frame*

- Cut a replacement piece out of material with the same or similar (oversize) section (Fig. 12.6b(iii)), or use squared unrebated section – the profile can be marked off the existing one after the joint has been fixed.
- Bore holes for the anchor peg and fixing screws (see Fig. 4.45). Treat all end grain and the back with wood preservative and – when dry – paint them.

- Place cement grout (sand and cement mix) around the peg hole. Coat the splice joint with waterproof mastic before tightening it by leverage from the floor and screw through the splice.
- Dress the joint with smoothing and rebate planes and make good the cement work around the base (Fig. 12.6b(iv)).

12.4.2.1 Loose door frames/jambs

These were originally secured back to the wall using steel hold-fasts as shown in Fig. 12.7a. These would be fixed to the face of the jamp and concealed with a cover lath.

However, 'frame fixings' (Fig. 12.7b), are now used as a modern alternative and, depending on the situation, a number of different varieties are available.

12.4.3 Sash repairs

See Table 12.3.

Table 12.3   Remedial treatment for defective sashes

Defect	Possible cause	Remedial treatment
Sticking	As doors (Table 12.2).	As doors (Table 12.2).
Sagging	Defective or worn hinge(s). Inadequate or defective setting or locating glazing blocks. Defective corner – rotten or damaged.	Renew hinge(s). Re-glaze and reposition glazing blocks as shown in Fig. 3.47. Repair* or remake sash.
Binding (springing)	As doors.	As doors.
Twisted	Construction fault; twisted frame.	Small amounts may be taken up by stays and fasteners – see also Fig. 12.4.
Rotten or damaged members	Wet rot – see Volume 1, Chapter 2.	Opening sashes should be remade. Fixed sashes could be repaired (spliced and false tenoned) – see Fig. 12.5.
Broken glass	Sash sticking; accidental, or criminal damage.	See Section 12.4.5 Re-glazing a window.

* Steel mending plates (angle plates) should only be used as a temporary measure

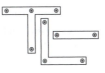

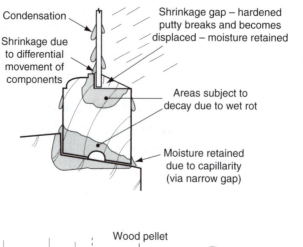

Fig. 12.8   *Decay as a result of permanent moisture presence*

Condensation

Shrinkage due to differential movement of components

Shrinkage gap – hardened putty breaks and becomes displaced – moisture retained

Areas subject to decay due to wet rot

Moisture retained due to capillarity (via narrow gap)

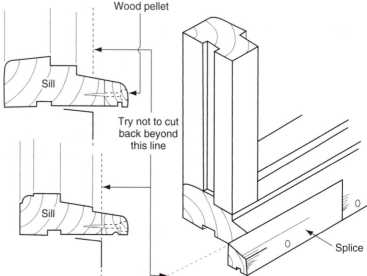

Wood pellet

Sill

Try not to cut back beyond this line

Sill

Splice

Fig. 12.9   *Repairing a damaged or rotten sill*

12.4.4 Window frame repairs

Note: *The removal of window frames is similar as for door frames. See Fig. 12.6a*

The sill and the bottom of jambs and mullions are the most likely places in which to find rot. Jambs and mullions can be dealt with in the same way as door jambs (Fig. 12.7). However, a word of warning – if the jambs or mullions form part of a bay window, it is highly likely that they have been used as loadbearing members to the floor or (more likely) the roof above in which case, before they are disturbed, all dead and imposed (live) loads bearing on the members must be suitably and safely supported (see Chapter 11). Defective load-bearing members should not be spliced but should be replaced as a whole.

Window sills can present problems, especially if the defects extend from outside to beyond the front line of the sash. In this case it would be advisable to replace the whole window (for example, see Volume 1, Fig. 2.10). Figure 12.9 shows how the nosing of a sill can be spliced.

12.4.5 Re-glazing a window

Work on large window panes should only be undertaken by specialist glaziers who are equipped for such work.

Where practicable, an opening light should be unscrewed from its frame and broken glass removed safely at ground level. However, broken glass within a fixed light must be treated with extra caution – operatives should at all times position themselves at a safe distance above and away from the window, in case any broken pieces of glass become dislodged during the work. The area below the window must be cordoned off with adequate warning signs displayed to ensure that no one enters this potential danger zone.

Before any work begins, operatives must be wearing either safety glasses or goggles, which must remain in position during the whole operation as protection from glass splinters which will inevitably be shed in varying degrees while the broken glass pane is being removed. Stout leather gauntlet-type gloves with rubber finger-grips should be worn to protect hands and wrists while handling glass – particularly when removing broken pieces.

12.4.6 Removing the broken glass pane

Starting at the head or top rail, work around the frame or sash removing old putty fronting (pointing) with a 'hacking knife' and hammer. After all the edges of the glass within the rebate have been exposed, remove any glazing sprigs (or panel pins) with pliers; then, using an old wide wood chisel held flat against the glass, gently tap the putty bedding with its blade as shown in Fig. 12.10. This action will dislodge the broken glass pieces from the rebate.

Repeat as necessary. Detached sashes can be laid flat on the ground during the latter part of this process. When all the broken pieces have been removed, old putty back bedding should be hacked away to leave a clean rebate that should then be painted (sealed) before re-glazing.

Before any glass is replaced or newly installed, consideration must be given to areas known as 'critical' glazed areas (see Chapter 4, section 4.6.6) and the use of safety glass (toughened glass). These are situations where there is a possible danger of breakage not only accidentally, but also by potential intruders.

Methods of glazing the window are explained in Chapter 3, section 3.8.

Note: *Any form of glazed areas can be a potential safety hazard whether installing or removing glass if not carried out correctly – using risk assessment procedures*

If the glass is 'beaded'-in, the operation as a whole is made easier, but beware of displacing pieces of glass when removing beads.

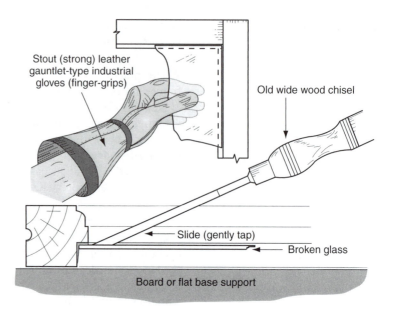

Stout (strong) leather gauntlet-type industrial gloves (finger-grips)

Old wide wood chisel

Slide (gently tap)

Broken glass

Board or flat base support

Fig. 12.10 *Dislodging broken glass from its framework*

12.4.7 Renewing sash cords

It is false economy not to renew all four cords if one sash cord breaks. There are many schools of thought as to how this operation should be carried out, but the method below will prove to be both quick and effective. All work is carried out from inside the dwelling, therefore the surrounding floor area should be protected. If a dust-sheet is used for this purpose, it should not be placed on a polished floor where you might stand or walk, as it will very likely slip from under you.

The following should be read in-conjunction with Fig. 12.11.

(i) Close and securely fasten the sashes to prevent sliding or falling over when beading is removed.

(ii) Cut any remaining cords to the lowered sash and gently lower the sash weight to the bottom of the case. (Never allow a weight to drop under its own weight, as this could damage the case and its fixing.)

(iii) Leaving the head and sill staff beads in place, remove the right- and left-hand staff beads (Fig. 3.18) by first using a sharp trimming knife to cut through any layers of paint covering the joint, then (after removing any screws) easing the beads out from the centre by using an old wide wood chisel or flooring chisel (see Fig. 12.12a) until they are bow-shaped and can be disengaged from their corner mitres. Nails should be pulled through the beads with pincers to avoid damaging the paint film. (Round-headed nails should not have been used but, if they have been, cut off the protruding portion with a junior hacksaw). The beads should then be laid at their respective sides of the window after marking (T) at their top ends.

(iv) Unfasten the lower sash. Lift it out of the window, remove old cord and nails from the stiles, and put it to one side.

(v) Unfasten and lower the top sash and cut its cords as at stage (ii).

(vi) Holding the sash steady to prevent it falling, remove the parting beads (after cutting through any paintwork as previously described) by starting at the sill end and easing them out of their grooves with a chisel. They may or may not be nailed – if they are, remove with care.

(vii) Lift out the top sash, remove old cords and nails from the stiles, and put it to one side.

(viii) Remove the pockets (one screw each side, Fig. 3.20) and take out the weights from their cases. Each weight should have its old cord removed and be chalk-marked 'F' (front) or 'B' (back), then laid on the floor at its appropriate side of the window.

Note: *Front and back weights may differ in size and/or weight*

(ix) Mark line (sash-cord) lengths as shown in Fig. 12.11a. Suitable distances X and Y, to allow for cord fixing and pulley clearance (Fig. 12.11e), are marked with chalk on the stiles of both sashes. These distances are then transferred to the face of the pulley stiles.

(x) Figure 12.11b shows how a cording 'mouse' can be made from either small 'rolled' lead sheet or a short length of sash chain (sometimes used on heavy sashes in place of cord). The mouse string (not less than 2 m long) is attached to the end of a 'knot' (hank) of good-quality waxed or recommended alternative sash cord as shown, and the mouse is made ready to be threaded over the pulley wheel. (Ensure that the pulley wheel is freely running and not broken – lubricate or replace as necessary.)

(xi) Figure 12.11c shows two alternative methods of cording the window. At the end of the run, the cord is attached to the appropriate back top sash weight (TSW) (see Fig. 3.23).

(xii) Working from the right-hand or left-hand side (depending on the method of cording), feed the corded weight for the top sash into the box (Fig. 12.11c). With the weight resting in the vertical position off the box bottom (making sure it is on the correct side of the parting slip, (Fig. 3.19) pull the weight up to just below the pulley and wedge the cord with a dowel peg. Then pulling the cord down the pulley stile, cut it off 50 mm above the chalk line X. Tie the bottom-sash weight (BSW) to the loose cut end of the cord, pull it up to the pulley and wedge it in that position, then cut the cord 30 mm below the chalk line Y. Repeat these processes on the opposite side. Fix the pockets back into position.

(xiii) Attach the cords to the sashes either by nailing with sash nails or by knotting, as shown in Fig. 12.11e (see also 3.23). Use at least four nails per stile – each nail should be slightly sloping away from the rebate to avoid touching the glass. Nails should not interfere with the pulley. Cord the top sash first, then release its pulley wedges and reposition the parting beads. Then cord the bottom sash. Replace the staff beads – if nails are to be used (screws in cups are preferred), use ovals.

(xiv) When the window is re-corded in this way, it should be noticed from Fig. 12.11f that when the top sash is closed its weights

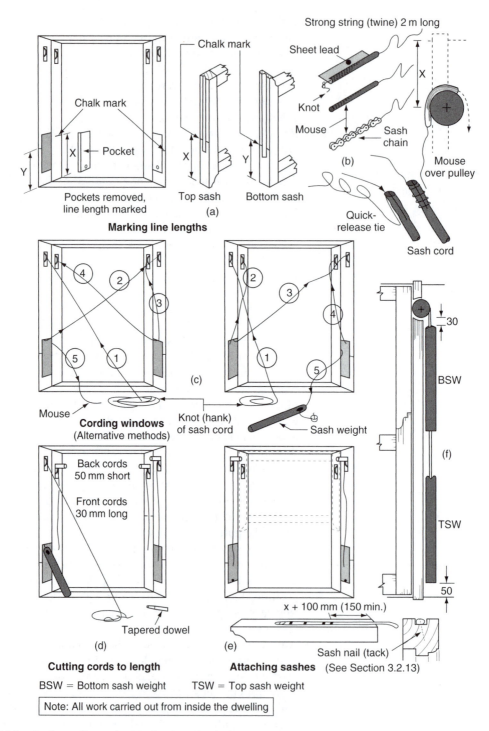

Strong string (twine) 2 m long

Sheet lead

Chalk mark

Chalk mark

X — Pocket

Y

Pockets removed,
line length marked

Top sash

Bottom sash

Knot

Mouse

Sash
chain

(b)

Mouse
over pulley

Quick-
release tie

Sash cord

Marking line lengths

(a)

Mouse

Cording windows
(Alternative methods)

Knot (hank)
of sash cord

Sash weight

(c)

30

BSW

(f)

TSW

50

Back cords
50 mm short

Front cords
30 mm long

Tapered dowel

(d)

(e)

x + 100 mm (150 min.)

Sash nail (tack) (See Section 3.2.13)

Cutting cords to length

Attaching sashes

BSW = Bottom sash weight   TSW = Top sash weight

Note: All work carried out from inside the dwelling

Fig. 12.11   *Sash-cording a double-hung sash window*

will be suspended 50 mm above the sill, thus allowing for any stretch in the cords, or build-up of mortar dust, etc., in the case bottom which could possibly interfere with the distance the weight can travel to allow the sash to close properly. The bottom-sash weights should be a minimum of 30 mm below the pulley when the bottom sash is closed. Ensure that the sashes run freely – rubbing candle wax over the sliding parts of the sash can ease their movement.

(xv) Remove dust from the sill and window board with a dust brush. Collect up spent nails and old sash cord, together with the dust sheets, and remove them from site.

12.4.8 Repairing floorboards

As mentioned in Table 12.1, the defect may be due to decay (rot). However, if it is due to damage caused by carelessness in forming an under-floor access trap (Fig. 12.12a), the two solutions are shown in Fig. 12.12(b) and (c): either simply replace the damaged board and provide bearings and joint cover as necessary or – if the damage covers more than one board – extend the repair or trap.

Warning: if a board has to be cut out of the floor and its underside is not visible, there is always the possibility that services, i.e. electric cables or water and gas pipes, may run beneath it. First switch off the mains electricity supply to the premises or check with adequate electric metal detector – it is better to be safe than sorry! – and proceed with caution.

Figure 12.12d shows how board ends can be cross cut. Damaged boards can be chopped out on to a joist, using a chisel, or cut alongside.

A jig saw is useful here, but only if the underside of the board is visible and clear of services. Alternatively, a series of small holes can be bored to help the point of a flooring saw or tenon saw to get started on piercing the board.

A pad saw can then be used to lengthen the cut and detect any pipework. (Flooring saws may differ in shape but are all designed to start a cut from off the surface of a board and to cut into corners. They are a most useful addition to the tool kit.)

A similar procedure can be used to rip down the tongue of a laid board. Equally useful is a small portable circular saw set so that its blade cuts the tongue but does not protrude below the underside of the board.

To remove or lift a board, find a suitable end joint and cross cut the other end of the length to be removed. Rip down one tongue if the tongue-and-groove joints between the boards are wide, or both tongues if close. Punch the nails to at least two-thirds of the board's depth. Using two wide flooring chisels (sometimes called 'tonguing tools' – *not brick layer's bolsters*), lever the board up from one edge.

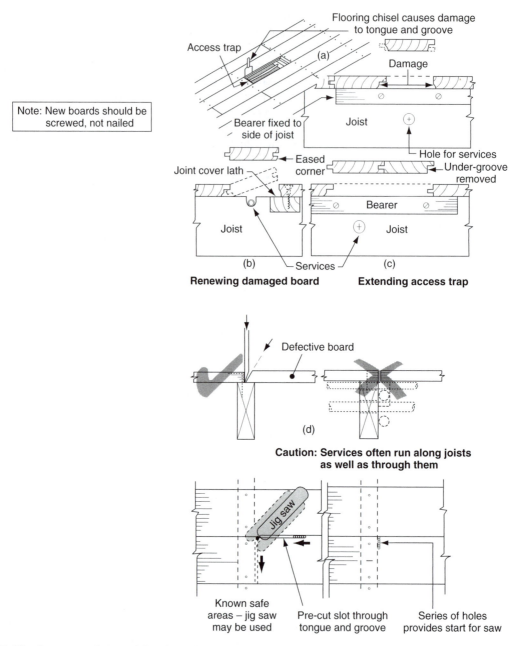

Note: New boards should be screwed, not nailed

Renewing damaged board

Extending access trap

Caution: Services often run along joists as well as through them

Fig. 12.12   *Renewing damaged floorboards*

12.4.9 Repairing stairs

Creaking treads and damaged nosings are among the most common faults associated with stairs. Access to the whole of a tread-and-riser staircase is usually not possible, due to its underside (soffit) having been plastered over.

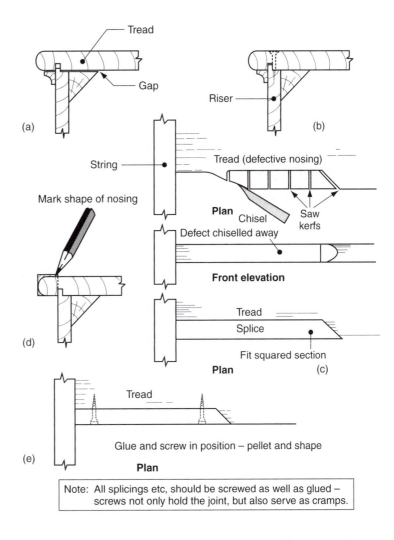

Fig. 12.13 *Surface repairs to a close-string stair*

Note: All splicings etc, should be screwed as well as glued –
screws not only hold the joint, but also serve as cramps.

A creaking tread is generally the result of the parting of the joint between the riser and tread (Fig. 12.13a). This can usually be remedied by gluing and screwing the tread on to the riser as shown in Fig. 12.13b. (If the treads are covered with carpet, the countersunk screw heads will not be visible; otherwise, wood pellets can be used over the screw heads.)

If this method does not work, then access to the underside of the treads will be necessary to check on glue-blocking, string housing, and mid-carriage support (if used). Adding steel brackets might be a possible remedy.

A defective nosing can be repaired by carefully cutting away the damaged portion and splicing on a new one (Fig. 12.13c). If possible, do not cut back beyond the riser face. Cut the splice end at 45° (avoid the walking line 300 mm in from either side) to lengthen the joint glue line. Fit the new rectangular section, then mark and shape the nosing profile (Fig. 12.13d). Glue and screw the new section into position, plug the screw-head holes with pellets, and sand to the finished shape (Fig. 12.13e). If a scotia or similar mould has been used between the underside of the nosing and the riser, renew the whole length – glue and pin it into position.

12.5 Door and window hardware

The working mechanism of locks, fasteners, bolts, etc. should be kept free from paint, etc. and be lubricated to ensure trouble-free use. (A dry graphite lubricant should be used for locks.)

Items of hardware found to be defective through wear should be replaced with their nearest alternative. It is possible, however, that the position of the new fixing holes, etc. will differ from the original ones, in which case the old screw holes can be plugged with glued wood dowel or wood filler. Unused holes (spindle holes, keyholes, etc.) and recesses should be concealed with a filling of wood as shown in Figs 12.14a–f.

The filling piece is cut first, with slightly tapered edges to self-tighten when fitting, placed on the timber and marked around the edges with a sharp knife before carefully chiselling out. Alternatively, to disguise a repair in timbers that are to receive a polish or varnish finish, a diamond shaped filling piece, of similar timber, may be used (as shown in Fig. 12.14(e) and 12.14(f)) which matches the grain and colour of the existing timber.

The lock keep will similarly have to be filled, as shown in Fig. 12.14(g) and (h).

12.6 Making good repaired areas

Making good small areas of brickwork, plasterwork and wall and floor tiles can be found in *Work Activities Section 13.5.3–5 by Brian Porter and Reg Rose*.

12.7 Health and safety

Before any work is carried out, particularly on older buildings, those responsible for the building must inform those people concerned of the location of *any* suspicious materials. Furthermore, a risk assessment must be undertaken to identify *any* dangerous situations including the following:

(a) Use of scaffolding (see Volume 2, Chapter 10);
(b) Unsafe building (see Chapter 11, shoring);
(c) Welfare of operatives and general public (see Volume 2, Chapter 2, fencing and hoardings);

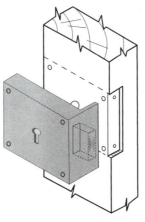

(a) Old rim lock – holes and recesses to be filled

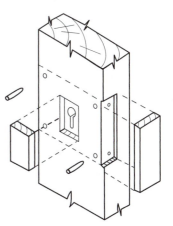

(b) Holes and recesses enlarged to receive filling pieces

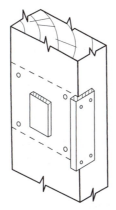

(c) Overthick filling piece glued and pinned in place

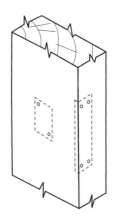

(d) Dressed flush with smoothing plane, fill nail holes and small gaps with wood filler – sand smooth

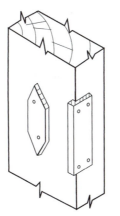

(e) Alternative method using diamond shaped filling piece glued and pinned in place (for polished timber to disguise repair)

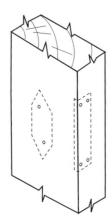

(f) Dressed flush with smoothing plane, fill nail holes and small gaps with wood filler – sand smooth

Fig. 12.14 *Making-good holes and recesses*

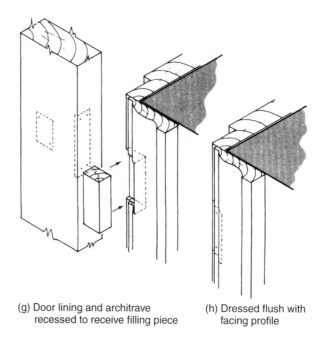

(g) Door lining and architrave
    recessed to receive filling piece

(h) Dressed flush with
    facing profile

Fig. 12.14   *(Continued)*

(d) Use of plant equipment, power tools (see Volume 1, Chapters 5 and 6);
(e) Environment (wildlife – including protected species, etc.). Restrictions regarding burning of discarded materials.
(f) Adjoining properties.

Additionally, harmful materials may be present and may be harmful not only to the operatives and the public, but also to the environment as a whole.

Some examples include:

(a) lead-based paints (lead poisoning due to burning off or sanding down);
(b) lead pipe-work (lead poisoning – handling);
(c) asbestos-based materials (which were common up to early 1980s before current restrictive regulations were introduced).

These could be found as either rigid materials for example

 (i) wall coverings (fire proofing);
 (ii) roof slates;
(iii) soffits;
 (iv) drain pipes;
  (v) cladding;
 (vi) corrugated sheet (roof and wall coverings).

or semi-rigid materials such as

(i) caulking (flexible jointing material used for sealing);
(ii) artexing (walls and ceilings) – since 1986 now contain asbestos substitutes. However, always check before work is carried out;
(iii) insulation.

In the case of (c) asbestos based materials or suspitious materials these must be reported, i.e. to the Health and Safety Executive representative, and the area sealed off before any work is carried out. If identified as dangerous, a specialist company must be employed to remove any material of this kind before work commences.

Note: *Suitable protective clothing must be worn at all times during general maintenance work even if there is only a slight suspicion of any dangerous materials (see section 12.3q)*

References

Health and Safety at Work Act (HSAWA) 1974.

Control of Substances Hazardous to Health Act (COSHH) 1994.

The Construction (Health, Safety and Welfare Regulations) 1996.

Building Regulation Approved Document N: 2000 Glazing – safety in relation to impact, opening and workmanship:

HSE (Health & Safety Executive) Safety leaflet 'Working with asbestos'.

HSE (1995) Asbestos alert for building Maintenance, repair and refurbishment workers be awareof asbestos the hidden killet. Pocket card INDG 188.

HSE Information Sheet General Access, Scaffolds and Ladders, Information Sheet 49.

BS 8417: 2003, Preservation of timber – recommendations.

BS 5973: 1993, Code of practice for access and working scaffolds and special scaffold structures in steel.

BS 1139–3: 1994, Specification for prefabricated access and working towers.

BS 1139–4: 1982, Specification for prefabricated steel splitheads and trestles.

BS 1129: 1990, Timber ladders, steps, tretles, and lightweight stagings.

BS 2037: 1994, Aluminium ladders, steps, and trestles for the building and civil engineering industries.

BS 2482: 1981, Specification for timber scaffold boards.

BS 5973: 1993, Section 9–tying to buildings.

Volumes and Chapters

Index